Volume 5

THE SPECIAL THEORY OF RELATIVITY

BOUND WITH

RELATIVITY

THE SPECIAL THEORY OF RELATIVITY

HERBERT DINGLE

BOUND WITH

RELATIVITY

A Very Elementary Exposition

OLIVER LODGE

Routledge
Taylor & Francis Group

LONDON AND NEW YORK

The Special Theory of Relativity first published in 1940
Second Edition published in 1946

Relativity first published in 1925
Second Edition published in 1925

This edition first published in 2014
by Routledge
2 Park Square, Milton Park, Abingdon, Oxon, OX14 4RN

and by Routledge
711 Third Avenue, New York, NY 10017

Routledge is an imprint of the Taylor & Francis Group, an informa business

First issued in paperback 2016

British Library Cataloguing in Publication Data
A catalogue record for this book is available from the British Library

ISBN13: 978-1-138-01351-3 (hbk)
ISBN13: 978-1-138-99003-6 (pbk)

Publisher's Note
The publisher has gone to great lengths to ensure the quality of this book but
points out that some imperfections from the original may be apparent.

Disclaimer
The publisher has made every effort to trace copyright holders and would
welcome correspondence from those they have been unable to trace.

THE SPECIAL THEORY
OF RELATIVITY

by

HERBERT DINGLE,
D.Sc., D.I.C., A.R.C.S.

**PROFESSOR OF NATURAL PHILOSOPHY
IMPERIAL COLLEGE OF SCIENCE AND TECHNOLOGY**

SECOND EDITION

METHUEN & CO. LTD.
36 ESSEX STREET. W.C.
London

First published . . . . . *November 10th, 1940*
Second Edition *1946*

PREFACE

THIS book is based on a short course of lectures given in recent years to Honours Physics students at the Imperial College.

So much has been written on the Special Theory of Relativity that a fresh book on the subject calls for some explanation. The apology offered here is that, so far as I know, the treatment adopted—namely, the development of the whole form of the theory from a re-definition, along ordinary scientific lines, of the measurement of length—has not previously been given. Such a treatment exonerates the theory from the last suspicion of metaphysics, and seems to me to present it in the form in which its significance for both science and philosophy can best be appreciated. The impulse to publish has been greatly stimulated by a recent correspondence in *Nature*, from which it is clear that there is still a deep and widespread disagreement on fundamental points of the theory. I hope the systematic development given here will enable the points at issue to be definitely located, and the truth, whatever it may be, brought to light.

A word should be said on the relation of this small work to the excellent little book on *Relativity Physics* by Professor McCrea, published as one of these Monographs. Apart from the characteristic just mentioned, the present treatment differs from McCrea's in two respects. In the first place it examines in detail the effect of the relativity theory on the fundamental concepts of physics, without attempting to estimate the influence of the resulting modification of physical concepts on the whole structure of physical theory. McCrea, on the other hand, is concerned mainly with the scope of the theory. The result is that the whole of this book is concerned with what McCrea compresses into a few pages, and from this point of view the book may serve as an introduction to his.

Secondly (rather to exaggerate a contrast in order to make it clearer), whereas McCrea's tendency is to present the world of experience as an exemplification of mathematical formulae, the theory is presented here as a mathematical formulation of relations first discovered in the world oi experience. For example, McCrea deduces the Fitzgerald ' contraction ' from the Lorentz transformation formulae, while here the transformation formulae are deduced from the Fitzgerald ' contraction '. Properly understood, of course, there is no incompatibility between the two procedures, and it may be of advantage to the student to have both available.

HERBERT DINGLE

May, 1940

CONTENTS

CHAP.   PAGE

I   RELATIVITY AND PHYSICAL PRINCIPLES   I

II   THE EXPERIMENTAL BASIS   10

III   RELATIVITY AND LENGTH   23

IV   TIME   37

V   VELOCITY AND ACCELERATION   53

VI   MASS, ENERGY AND FORCE   62

VII   ELECTROMAGNETIC MEASUREMENTS   76

VIII   TRANSITION TO GENERAL RELATIVITY   81

INDEX   93

CHAPTER I

RELATIVITY AND PHYSICAL PRINCIPLES

The Fundamental Principle. The Principle of Relativity may be stated thus : *There is no meaning in absolute motion.*

By *absolute* motion is meant motion assignable to a body or to any locatable phenomenon (e.g. a spot of light, region of high temperature, &c.) without reference to some other body or locatable phenomenon. By saying that such motion has *no meaning*, we assert that there is no observable effect by which we can determine whether an object is absolutely at rest or in motion, or whether it is moving with one velocity or another.

Thus, the statements that the Sun is at rest, that the Earth is moving, and that the velocity of the Moon's shadow during a solar eclipse is 2,000 miles an hour, are all, taken literally, statements concerning absolute motion. They are without meaning, according to the principle of relativity, because nothing observable would be different from what it is if we supposed any or all of them to be either true or false.

On the other hand, statements concerning *relative* motion have meaning, and are either true or false. Thus, the statement that the Sun is at rest with respect to the mean position of the surrounding stars is a meaningful statement which is false ; the statement that the Earth is moving with respect to the Sun is a meaningful statement which is true ; and the statement that the velocity of the Moon's shadow over the surface of the Earth during a

solar eclipse is 2,000 miles an hour is a meaningful statement which may be either true or false.   The reason why
these statements have meaning is that if we change the
motion expressed in any of them (e.g. if we state that the
velocity of the Moon's shadow over the Earth in a particular eclipse is 2,500 miles an hour) a change in certain
observable phenomena is thereby implied, and these phenomena could therefore be used to determine which is the
false statement.

It follows from the principle of relativity that there is
no meaning in speaking of the motion of the whole observable universe, for there can be nothing outside to which
to relate such motion.   In other words, if we suppose the
whole universe to be suddenly moved with any speed in
any direction, then, provided that all the *relative* motions
within it are unchanged, no one could know of the fact.
It would, however, be unscientific to suppose that something had happened, of the occurrence of which we could
not, in the nature of things, become aware.   According to
the scientific tradition, experience is our starting-point, not
a more or less accidental awareness of some independently
existing reality : we should therefore say that *nothing* had
happened—i.e. that the assertion of such a change of
motion was a statement without meaning, akin to the
statement that the universe had suddenly become tradified.
If, in course of time, someone discovers an effect by which
an absolute motion can be assigned to a body, our criterion
of meaning will not be altered : we shall simply say that
the principle of relativity is false.

At present, however, we believe it to be true—not only,
or primarily, because it appeals to our common sense, but
because the whole of experience of which we have any
record contains no evidence of any observable effect of
absolute motion.   What appear to be such effects invariably
turn out on examination to depend on relative motion
only.   Let us take one or two examples.

Consider a circular coil of wire, and a magnet lying along its axis at some distance from the centre and at rest with respect to the coil. Let a galvanometer be included in the coil. So long as the arrangement remains as described the galvanometer needle will remain at the zero point. Now move the magnet towards the coil. While the motion lasts the needle will be deflected, and when it ceases the needle will return to the zero point. Here is apparently a criterion of motion—the magnet is moving if the needle is deflected. Actually, however, the deflection indicates not the absolute motion of the magnet but the relative motion of the magnet and coil. If the magnet is moved exactly as before and the coil also moved with the same speed and in the same direction, no deflexion is observed. Further, precisely the same deflexion as before may be produced by keeping the magnet still and moving the coil towards it, if the relative velocity and distances are kept the same.

Another example is given by the Doppler effect. Consider a spectroscope and a distant source of monochromatic light. A spectrum line is observed in a certain position. If, now, the source moves towards or away from the spectroscope, the line is displaced to one side or the other. Again we have apparently an effect of the motion of the source of light, and again it is merely an effect of the relative motion of the source of light and something else— namely, the spectroscope. For if the spectroscope be moved with the same speed and in the same direction as the light, no displacement of the line is seen ; and, as before, if the source of light is kept stationary and the spectroscope moved towards or away from it, the displacement reappears.

The principle of relativity is a generalization from the fact that all known effects, apparently caused by the intrinsic motion of a single body, depend on the motion of that body with respect to another object. In this it resembles the second law of thermodynamics : the principle that heat cannot by itself pass from a cooler to a hotter body is likewise a generalization from the fact that in all our experience we have never observed such a thing to happen. Both principles are similarly vulnerable to observation : a single well-authenticated instance of a violation would suffice to overthrow either of them.   There is, however, a certain *a priori* probability about the principle of relativity which does not pertain to the second law of thermodynamics.   Apart from experience we could scarcely expect that if we brought two bodies into contact they would come to a state in which a thermometer stood at the same level when placed in each of them : it might equally well have been that, in certain circumstances, all the heat would leave the cooler and settle in the other.   But if the reader imagines himself alone in empty space, with no landmarks, winds, or any other external reference mark to tell him that he is moving, he will probably feel no difficulty in assenting to the proposition that it is immaterial whether he considers himself to be at rest or moving in any arbitrary manner, and that the onus of proving that absolute motion has meaning lies on him who makes the assertion. Whatever difficulty relativity holds for the student does not arise from the incredibility of the fundamental principle.

The Impact of Relativity on Physics.   It will be seen that the principle of relativity is purely negative in character : it asserts merely that ' absolute motion ' is a meaningless succession of letters.   The question naturally arises : how can such a principle lead to anything significant ?   ' Nothing can come of nothing.'   We should not expect an important branch of physics to originate in the recognition that ' the universe is tradified ' means nothing : why, then, should the case be altered when the meaningless symbol is given another form ?

The answer is that the case is not altered.   The principle of relativity in itself tells us nothing whatever about anything, and if we were now about to begin the physical description of the universe, and paid full regard to the principle, we should still have to make the fundamental experiments, form our elementary concepts through which to relate our observations together, and gradually build up the edifice of physics in the same general way as did Galileo, Newton, and their successors.

The fact is, however, that we are not now about to begin the physical description of the universe. We have already advanced a long way with that description, and what had been done up to the end of the nineteenth century had been done *without* full regard having been paid to the principle. The consequence is that we have inherited a complicated structure of physical conceptions and theories, the coherence of which depends on the possibility of physical effects arising from a change in the absolute motion of a body. No such effects have, of course, been observed, or the principle of relativity would not have been formulated; their persistence in not appearing was, in fact, one of the most puzzling and exasperating features of nineteenth-century physics. But their essential *observability* could not be denied without making necessary a thorough transformation of theoretical physics.

It is that transformation which constitutes the branch of science known as 'relativity theory'. It is not a trifling change, touching only the latest refinements of physics; it affects the very foundations, the primary definitions and concepts on which the whole science is built. Nevertheless, the defect which it has been called into existence to remove is such as to reveal itself conspicuously in only two classes of phenomena. We find ourselves, as it were, with an organism of physical theory on which we observe two patches of skin disease. Pursuing their origin, we find ourselves forced to penetrate to the very heart of the organism and remove from it that element of its nature which implies the possibility of observable effects of absolute motion.   Let us look at the skin diseases.

The first has to do with the motion of matter through ether.   In order to explain the fact that we see distant bodies through transparent matter we postulate the existence of something called 'light' which consists of waves in a universal 'ether' filling all space and penetrating between the atoms of bodies.   To this light we assign, on the basis of certain experiments, a definite velocity with respect to the ether of about 3×10^{10} cm. a second in empty space—a quantity universally represented by the symbol c.   The ether itself is unobservable, but it is essential to the whole conception that a definite location in space shall be assignable to each element of it, so that we can say that at a certain instant the wave-front of a beam of light is at such and such a place in it, and one second later it is 3×10^{10} cm. further on in a definite direction.   It follows that we can speak significantly of the velocity of a body with respect to the ether, that velocity being measurable by dividing the distance between two specified elements of the ether by the time the body takes to go from one to the other.   We cannot carry out this process directly because we cannot specify particular elements of the ether by direct observation, but we can make an equivalent indirect experiment in which a beam of light is used to locate positions in the ether.   The velocity of light through the ether is c, and if we are moving through the ether with velocity v in the same direction as the light, the velocity of light relative to us will be $c - v$, whereas if we are moving in the opposite direction the velocity will be $c + v$.   If we are at rest, or moving in any other direction, the velocity will lie between these extremes.   To find v we have therefore to measure the velocity of light coming from various directions, choose the maximum and minimum values, and take half the difference between them : we shall then have obtained our velocity through the ether.

Now this is an absolute velocity of a single body—the experimenter—for the ether is unobservable. Moreover this absolute velocity has a meaning, for it affects the value we obtain for the velocity of light. Hence the nineteenth-century theory of the ether is inconsistent with the principle of relativity. This revealed itself as a defect in theoretical physics when it was found by trial that v was by no means discoverable in this way or in any other that could be thought of.

The second 'skin disease' is associated with the conception of 'fields of force'—in particular, gravitational fields of force. According to Newton's law, the natural state of a body not subjected to a force is that of rest or uniform motion in a straight line : if a force acts on it, however, the body becomes accelerated. Now gravitational force is conceived to be a 'physical reality'—something potentially capable of detection apart from the acceleration, which is merely its effect on a material body which happens to be a victim of it. It must be of the same character as the impact of visible, tangible matter, for an object resting on a table is conceived to be in equilibrium through the neutralization of its (gravitational) weight by the resistance of the table. Suppose, then, that by some means other than its accelerating effect, we discover that a gravitational field of a certain strength, and no other force of any kind, exists in a certain region (no such means has ever been devised—this again is the outward sign of the disease— but that does not alter the possibility inherent in the conception). We can then calculate the acceleration it will produce in a material body. Let us then set free a body in the field, measure its acceleration with respect to our laboratory, say, and deduct the calculated effect of the field. The remainder must then be the absolute acceleration of the laboratory.

Notice an important difference between these two implications of the significance of absolute motion. The first allows a meaning to be given to absolute *velocity* : the second allows a meaning to be given only to absolute *acceleration*, for uniform velocity and rest are indiscriminately associated as the 'natural' state of a body unaffected by a force.   Our two skin diseases are therefore symptoms of two at least partially independent defects of constitution.   The remedy for the first is to insist on obedience to the *special* or *restricted* theory of relativity, viz. : *There is no meaning in absolute velocity*.   The remedy for the second is to insist on obedience to the *general* theory of relativity, viz. : *There is no meaning in absolute acceleration*.

In this book we shall be concerned only with the special theory, except for a brief note on the transition to the general theory.   It may be remarked at once, however, that it is only from a certain viewpoint—that of a particular mathematical formulation—that the special theory becomes a particular case of the general theory.   In essence the two theories are independent, and the modification of pre-relativity physics which we shall find to be required by the special theory cannot be deduced from the general theory in its present state.

The course which we shall follow may be indicated briefly as follows.   We shall consider a few of the more important experiments which, according to traditional physical ideas, might have been expected to lead to a knowledge of the absolute velocity of the Earth—i.e. the velocity of the Earth through the ether.   Their failure to do so will then be examined to see what error in traditional ideas led to the belief that they would succeed, and hence to find the correction which must be made.   We shall find that this correction can be most simply expressed as a substitution, for what we have been accustomed to call the 'length' of a body, of a slightly more complicated expression.   This substitution must be made whenever length occurs, explicitly or implicitly, in a physical relation, and since every physical measurement that is made depends in part on measurement of length, all physical measurements are thereby affected. The main part of our work is then to determine, from consideration of the classical definitions of the various physical quantities—time, mass, velocity, force, &c.—all the necessary corrections to be applied to them.

The subject known as ' the special theory of relativity ' is thus a revision of the basic principles of physics. In pursuing it we retrace the course followed in the past by those who established the subject, inserting at each step the modifications resulting from a slight change at the beginning. In doing so we obtain a clearer insight into the essential principles of physics, and become aware of possibilities which have always been inherent in the fundamental definitions but have remained unsuspected.

CHAPTER II

THE EXPERIMENTAL BASIS

MANY unsuccessful attempts have been made to observe the velocity of the Earth through the ether. We shall discuss two of the most important experiments bearing on the question—the Fizeau experiment and the Michelson-Morley experiment (repeated later in a modified form by Kennedy and Thorndike) and leave it to be understood that these are merely outstanding representatives of many experiments which together constitute the observational basis of the special theory of relativity.

The Fizeau Experiment. Refraction of light by matter is caused by a reduction in velocity, so that light takes longer to travel a given distance through matter than in a vacuum.

It must be understood that light, whether in empty space or in matter, is conceived as a train of vibrations in the ether. The fact that its velocity is less in matter than in empty space must therefore be attributed to some modification of the ether by matter : light itself is not to be thought of as waves in matter.

Consider a piece of transparent material—say a rectangular glass block—of thickness d, on which a horizontal beam of light falls normally. The block—supposed at rest on the laboratory table—is being carried with the earth through the ether at an unknown speed in an unknown direction. Let the component of its velocity in the direction opposite to that of the light be v. Then, choosing the laboratory as our standard of rest, the block may be regarded as situated in a stream of ether which, if it were unaffected by the air and glass through which it passes, would travel through them with velocity v in the direction of the light. We know, however, that ether is not unaffected by matter, so let us suppose that the velocities of the ether-stream through air and glass, respectively, are $\alpha_1 v$ and $\alpha_2 v$, where α_1 and α_2 are fractions depending on the unknown degree in which the two materials may retard the motion of ether.

So much for the ether : now consider the light travelling through it. Let the velocity of light in air relative to the ether in air be V_1, and the velocity of light in glass relative to the ether in glass be V_2. We shall then have, for the refractive index, μ, from air to glass,

$$\mu = \frac{V_1}{V_2}. \qquad \qquad \cdots \cdots \quad (1)$$

Also, the velocities of light in air and glass relative to the air and glass will be respectively, $V_1 + \alpha_1 v$ and $V_2 + \alpha_2 v$, so that the time-retardation which the light suffers by having to pass through the glass will be

$$\frac{d}{V_2 + \alpha_2 v} - \frac{d}{V_1 + \alpha_1 v}, \qquad \cdots \cdots \quad (2)$$

which, on expansion in powers of v, becomes

$$d\left[\left(\frac{1}{V_2} - \frac{1}{V_1}\right) - v\left(\frac{\alpha_2}{V_2^2} - \frac{\alpha_1}{V_1^2}\right) + \text{higher powers of } v \ldots \right] (3)$$

Now this is a function of v, and therefore, if we measure the retardation of the light for various orientations of the block (the light always being normal to it), we should be able to determine, from the position for maximum or minimum retardation, the direction and magnitude of the Earth's motion through the ether.

This procedure was suggested in principle by Arago early in the nineteenth century. On performing an equivalent experiment, however, he found that the retardation was

the same for all orientations of the glass.   The explanation
was given by Fresnel, who pointed out, in effect, that if
α_1 and α_2 were such as to satisfy the condition

$$\frac{\alpha_2}{V_2{}^2} - \frac{\alpha_1}{V_1{}^2} = 0, \quad \cdots \cdots \quad (4)$$

the term in v would vanish, while the terms in higher
powers of v might be too small to be detected.   In that
case the retardation would be simply

$$\frac{d}{V_2} - \frac{d}{V_1}, \quad \cdots \cdots \quad (5)$$

exactly as though there were no ether stream through the
laboratory : in other words, the phenomenon of refraction
would afford no information concerning the motion or rest
of the Earth with respect to the ether.

The reality of condition (4) may be tested in the following way.   So far we have supposed the glass to be at rest
in the laboratory : now let us suppose it moves with
velocity u (relative to the laboratory) towards the incident
light.   Its velocity with respect to the free ether will then
be $v + u$, and the velocity of the ether stream through the
glass relative to the glass will be $\alpha_2(v + u)$.   The glass,
however, is moving at velocity $- u$ with respect to the
laboratory, so that the velocity of the ether stream through
the glass with respect to the laboratory will be

$$\alpha_2 v + u(\alpha_2 - 1).$$

The velocity of light with respect to this ether stream remains V_2 : this we know from Arago's and later experiments which imply that the refractive index (see equation
(1)) is the same whether the glass is moving or at rest.
Hence the velocity of light with respect to the laboratory
will be $V_2 + \alpha_2 v + u(\alpha_2 - 1)$.

When the glass was stationary on the table, the velocity
of light in it with respect to the laboratory was the same
as that with respect to the glass—namely, $V_2 + \alpha_2 v$.   The

effect of the motion, then, is to increase the velocity of light with respect to the laboratory by $u(\alpha_2 - 1)$. Now if condition (4) is satisfied we may evaluate this in terms of observable quantities. Let us, for simplicity only, make the usual assumption that air is equivalent to a vacuum, so that $\alpha_1 = 1$.

Then (4) becomes

$$\left.\begin{aligned} \alpha_2 &= \frac{V_2^2}{V_1^2} \\ &= \frac{1}{\mu^2}, \text{ from (1).} \end{aligned}\right\} \quad . \quad . \quad . \quad . \quad (6)$$

Hence the increase in the velocity of light is

$$u\left(\frac{1}{\mu^2} - 1\right), \quad . \quad . \quad . \quad . \quad (7)$$

a quantity which can be measured.

An experiment to test this result, using water instead of glass, was first made by Fizeau. A beam of light was divided into two, which were sent in opposite directions round a path containing two parallel tubes filled with water, and afterwards brought together again in a telescope in which interference fringes were observed. At first the water was kept at rest, and the position of the fringes was noted. It was then made to travel along the tubes at a uniform velocity, u, which was measured, in such a way that the light in one beam travelled with the water, and that in the other against it. According to (7), the velocity of one beam should be increased, and that of the other decreased, by $u(1/\mu^2 - 1)$, so that the velocities of the beams should differ by $2u(1/\mu^2 - 1)$. This should cause the interference fringes to shift by a definite amount, and the shift would serve as a criterion of the change of velocity.

Fizeau's experiment—which has been repeated by others with greater precision and with a consistent result—gave the expected confirmation of the expression (7).

The aspect of this result which is of interest to us is that a method of finding the absolute velocity of the Earth which seemed to promise success was made of no avail because it happened that matter affected ether in just the manner necessary to ruin it. From the older point of view, this simply added detail to the ether theory. If we suppose light to consist of waves travelling through an ether which permeates matter, we must also suppose that moving matter exerts a partial drag on the ether in such a way as to give the same refraction phenomena whatever the speed of motion might be. Such an elaboration, however, comes perilously near to making nonsense of the ether theory, for since, by (6), α_2 is a function of μ, which itself is a function of the wavelength of the light, we must apparently have a separate ether to transmit each wave-length, and the drag on each of these ethers must be so adjusted to the corresponding wave-length as to vitiate our efforts to measure the velocity of the laboratory with respect to any of them. Such a result does not inspire confidence that a full understanding of the matter has been reached.

It may be remarked that condition (4) is sufficient to explain also the observation of Airy that the value found for the constant of aberration is the same whether the telescope used in measuring it is filled with air or with water.

The Michelson-Morley Experiment. The most famous of all experiments intended to measure the motion of the Earth through the ether is the Michelson-Morley experiment, first performed in an elementary form by Michelson alone in 1881, and repeated in improved forms several times later by various experimenters. It differs from the experiments previously mentioned in that the light used travels in the same medium throughout, so that no question of differential ether-drag arises.

In the Michelson-Morley experiment a beam of light

was divided into two parts which were made to travel along different paths in air and then re-united. The arrangement is shown in Fig. 1, in which the path of the light is shown in (*a*) on the supposition that the Earth and apparatus are at rest in the ether, and in (*b*) on the supposition that they are moving towards the right with uniform velocity. In (*a*), light from *L* falls at *A* on a half-silvered mirror which divides it into a beam *ABAT* and a beam *ACAT*, *B* and *C* being similar mirrors. These beams are

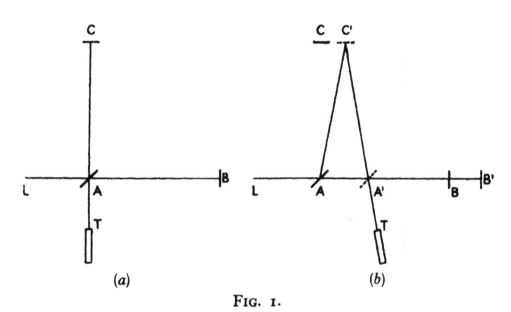

(*a*)   (*b*)

FIG. 1.

re-united along *AT* and produce interference fringes in the telescope *T*. The distances *AB* and *AC* were made the same, and it is clear that if the system were rotated in the plane corresponding to that of the diagram, there would be no change in the distances traversed by the beams of light, and therefore no change in the interference pattern.

   If the apparatus is moving in the ether, however, the case is different (Fig. 1 (*b*)). Here the distances traversed by the two beams are not the same, for during the passage of the light from *A* to a mirror and back again, the system has moved forward a certain distance. The path taken by one beam is now $AB'A'T$, and that taken by the other is $AC'A'T$, and it can easily be shown that these are not equal.

Let the velocity of light in air be c, and the velocity of the apparatus through the ether in the direction AB be v, and let $AB = AC = l$. Consider first the path $AB'A'$. The time which light takes to travel along this path is the same as though it had velocity $c - v$ along AB' and $c + v$ along $B'A'$, the total time being therefore

$$t = \frac{l}{c - v} + \frac{l}{c + v} = \frac{2lc}{c^2 - v^2} \quad \cdot \quad \cdot \quad \cdot \quad (8)$$

The actual velocity of the light, however, is a property of the ether only, and is equal to c. The distance the light covers in this time is therefore ct; i.e.

$$AB' + B'A' = \frac{2lc^2}{c^2 - v^2} = \frac{2l}{1 - v^2/c^2} \quad \cdot \quad \cdot \quad (9)$$

Now consider the path $AC'A'$. By symmetry this is twice AC', and

$$AC' = \sqrt{AC^2 + CC'^2}.$$

Now $AC = l$, and since the mirror travels from C to C' while the light travels from A to C', we must have $CC' = AC'.v/c$. Hence

$$AC' = \sqrt{l^2 + \frac{v^2}{c^2}AC'^2}\,; \quad \cdot \quad \cdot \quad \cdot \quad (10)$$

i.e. $$AC' + C'A' = 2AC' = \frac{2l}{\sqrt{1 - v^2/c^2}}. \quad \cdot \quad \cdot \quad (11)$$

Obviously (9) and (11) are not equal unless $v = 0$ or c, and the difference between them is a function of v. Hence, if now the system is rotated as before, a shift of the interference fringes should be observed. In the experiment, however, a shift was not observed.

The explanation that first suggests itself is that $v = 0$

(the possibility $v = c$, which would give equal infinite paths for the light, is ruled out by the fact that the light did return and enter the telescope). We can scarcely suppose, however, that the Earth is the one body in the universe which is permanently at rest in the ether, and we should therefore expect that if, at the moment of the experiment, v happened to have a zero value, it would have a different value six months later when the Earth's velocity with respect to the Sun would have changed by some 37 miles a second on account of its orbital motion. Repetition of the experiment at that time, however, gave the same result as before : the expected shift of the fringes was not observed.

Of the suggested explanations, some can be summarily disposed of. The idea that the velocity of light depended not on the ether alone but on the velocity of the material source of light also, could not be maintained because it would require a spurious eccentricity to appear in the orbits of double stars, which, as de Sitter showed, was not consistent with observation. Experiments are also fatal to the idea that the velocity of light is affected by reflection from a moving mirror. It is impossible, again, to suppose that the Earth by dragging gives its velocity to the neighbouring ether, because a contrary supposition is required to explain the aberration of light on the ether hypothesis. The only explanation that calls for consideration is that first suggested by Fitzgerald, and independently by Lorentz, that a material body moving through ether is automatically contracted by a factor $\sqrt{1 - v^2/c^2}$, in the direction in which the component of velocity is v. If that were so, the factor l in the expression (9) would become $l\sqrt{1 - v^2/c^2}$, and (9) and (11) would become equal for all orientations of the apparatus and for all values of v. The experiment would therefore necessarily fail to give any knowledge of the motion of the Earth through the ether.

Like Fresnel's explanation of Arago's experiment, this suggestion contains nothing exceptionable in itself. It is quite consistent with the electrical theory of matter, electrical and optical phenomena being correlated through the appearance of c as the velocity of light and also as the ratio of electrical units. It simply adds another detail to the theory of the interaction of matter and ether.

The Kennedy-Thorndike Experiment. The consistent failure of every experiment designed to provide evidence of the motion of matter through ether, associated with a number of independent ' discoveries ', each accounting for the failure of particular experiments, is a phenomenon which leads one to fear that if a new experiment were devised, some other compensating factor would be ' discovered ' to nullify it. In other words, we begin to suspect that there might be some fundamental reason why all attempts must fail. The theory of relativity says that there is such a reason, which is simply that what is looked for is meaningless. If that were so we might expect that some of the special explanations called forth by particular experiments might contradict one another, or that an experiment might be found whose failure defied explanation.

At the time when the theory of relativity was struggling to establish itself, this had not occurred, but in 1932 Kennedy and Thorndike announced the result of a slightly modified form of the Michelson-Morley experiment which appears inexplicable on the ether hypothesis. In this experiment the lengths AB and AC were made unequal, but sufficiently near to one another for good interference fringes to be observed. The result was again that no evidence of the motion of the Earth through the ether was obtained.

It is easy to see that in these circumstances the Fitzgerald-Lorentz hypothesis will not explain the result. For, if $AB = L$, while AC remains equal to l, the lengths of the light paths, after correction of the former for the proposed contraction, become

$$AB' + B'A' = \frac{2L}{\sqrt{1 - v^2/c^2}}; \quad . \quad . \quad . \quad (12)$$

and

$$AC' + C'A' = \frac{2l}{\sqrt{1 - v^2/c^2}}$$

and the difference—namely,

$$\frac{2(L - l)}{\sqrt{1 - v^2/c^2}}, \quad . \quad . \quad . \quad . \quad (13)$$

depends on both $(L - l)$ and v. Yet no evidence of the influence of either of these quantities was obtained. This leaves us with no explanation, consistent with the hypothesis of a localisable ether, of our inability to detect the motion of matter through it.

Descriptions of the Experiments. It has become customary to describe the Michelson-Morley experiment in terms of the times taken by the two beams of light to travel along their respective paths, instead of in terms of what was actually observed—namely, the interference pattern produced. For popular exposition this is perhaps good enough : it deals sufficiently accurately with the point at issue and introduces nothing that the ordinary non-scientific reader might not be expected to understand. For scientific purposes, however, such a description is most misleading, and has, in fact, led many into error. It is needless to point out that we do not measure the time of travel of the light at all, but it is very necessary to point out that the whole experiment is quite independent of any time-scale which we may adopt, and therefore quite inexplicable in terms of a postulated effect of motion on clocks or of our choice of equal time intervals.

The only measurements made in the experiments are of the lengths of the arms, their inclination to one another, and the position of the fringes. It is obvious that whether the watch in the observer's pocket runs fast or slow or keeps good time, the position of the fringes will be just the same.   It is true that if there had been a shift of the fringes, and we had then attempted to deduce from it what the value of v must be, we should have had to imply a particular system of time measurement; otherwise there would be no meaning in our statement of the velocity. But that is not the state of affairs.   There was no shift of the fringes, and it is impossible to explain that fact by any considerations at all concerning the measurement of time. It is true also that an approximation to a particular time scale is implied in the assumption that the velocity of the Earth was constant during the experiment (see p. 43), but that again cannot explain the constancy of the interference pattern; it merely enables us to state what kind of motion (namely, that usually called 'motion with constant velocity') the experiment has failed to detect.

This is particularly important in connexion with the Kennedy-Thorndike experiment, because it has been proposed to explain the null result obtained there by assuming, as well as the Fitzgerald contraction, a modification of the period of the atoms emitting the light, produced by their motion through the ether.   Such a hypothesis, however, cannot afford an explanation, because, whatever effect motion may have on the frequency of light, that effect must be shared by both beams equally: it cannot, therefore, be invoked to neutralize a shift which depends on a difference between the beams.

Miller's Experiments.   There is one possible exception to the statement that all attempts to measure the velocity of matter with respect to ether have failed.   Professor Dayton Miller, who was associated with Morley in earlier repetitions of the Michelson-Morley experiment, has conducted an elaborate series of observations with a refined form of the Michelson interferometer, and has obtained results which he interprets as indicating a slight but definite positive effect.*   He believes that this indicates not only the orbital motion of the Earth but also the motion of the Solar System through the ether, with a velocity of about 208 km./sec., towards a definite point in the sky.   The effect is much smaller than would be expected from such a velocity, and this Miller attributes either to a dragging of the ether by the Earth or else to the operation of a slightly modified Fitzgerald contraction. The slight displacements of the fringes observed by Michelson and Morley in their experiments, and attributed by them to experimental errors, are held by Miller to be genuine effects consistent with this hypothesis.

It is very difficult to explain these experiments. If Miller's results are accepted, we then have the problem of accounting for the various experiments which contradict it.   Miller has suggested that as several of these were conducted within thick closed walls, the dragging of the ether might have been increased sufficiently to reduce the effect below the limit of observation. This, however, would not apply to all of them, and if it is accepted we must presumably reject the modified Fitzgerald contraction and fall back on a dragged ether. We are then in difficulties with the aberration of light and other phenomena.

On the whole, therefore, the most legitimate attitude at present seems to be to attribute Miller's results to some unknown spurious disturbance. The position, however, is not so satisfactory as one would wish, since a theory which, like relativity, depends on a unanimous verdict of experiment, cannot admit exceptions. Nevertheless, there is no slackening of principle in tentatively ignoring these results. Where reports of observations are contradictory, some must be rejected, and we have to decide on which side the balance of evidence lies. If we accept Miller's results we do not merely destroy a theory ; we destroy

* See *Reviews of Modern Physics*, **5**, p. 203 (1933).

belief in the rationality of our experience, in what we used to call the 'uniformity of nature'. It is much to be hoped that a satisfactory explanation of the apparent contradiction will be found.

CHAPTER III

RELATIVITY AND LENGTH

The Definition of Length. On the theory of relativity, the experiments described in the last chapter need no explanation : if absolute velocity is meaningless, we do not need to account for our inability to measure it. But the matter cannot rest there. We are bound to ask why it was that we expected to be able to do so. There must have been some defect in our reasoning : what was it ?

The answer is very simple : it is that we have used a false definition of length. The quantity that is physically important is not l, as ordinarily defined, but $l\sqrt{1 - v^2/c^2}$, where v is the velocity of the object concerned in the direction in which the length is measured, with respect to whatever standard of rest we choose to adopt.

At first sight this may seem a very arbitrary statement. We shall presently try to show that it is not so, and even that it might in essence have been anticipated, but first let us see it in application to the Michelson-Morley experiment. The fact to be explained is that the position of the interference fringes is unaltered if the apparatus is rotated at any time of the year. Now the position of the fringes depends primarily on the difference in the number of light-waves in the paths of the two beams : the lengths of those paths are unimportant except in so far as they affect this difference. If, then, λ is the wave-length of the light used, and l the length of either path as ordinarily measured, we must now write $\lambda\sqrt{1 - v^2/c^2}$ and $l\sqrt{1 - v^2/c^2}$, respectively, for these quantities, so that the number of light waves in the path is

$$\frac{l\sqrt{1 - v^2/c^2}}{\lambda\sqrt{1 - v^2/c^2}} = \frac{l}{\lambda}$$

for all values of v—that is to say, in whatever position the apparatus may be at any time, and whatever arbitrary standard of rest we may choose. This applies equally to both paths, so that the difference in the number of light-waves in the two paths is always zero. Fig. 1 (b) then has no significance, for the motion which it connotes—motion with respect to the ether—is meaningless. The paths of the beams of light are always those shown in Fig. 1 (a).

In the Kennedy-Thorndike experiment we find similarly that the number of waves in one path is $\frac{L}{\lambda}$, and that in the other, $\frac{l}{\lambda}$, the difference being $\frac{L - l}{\lambda}$. This, again, is independent of v, so there is no shift of the fringes.

The difference between this and the hypothesis of the Fitzgerald contraction is clearly profound, although in many of their superficial aspects the two explanations seem almost identical. The contraction hypothesis can claim support from the electrical theory of matter, only for the shortening of material bodies with motion : that theory does not require the similar shortening of light waves. It cannot, therefore, explain the general Kennedy-Thorndike experiment, but only that particular form of it—the Michelson-Morley experiment—in which the wave-length of the light has no bearing on the result. On the relativity explanation we do not postulate a motion through the ether which is compensated for by another physical effect. We regard such a motion as a purely conceptual matter, depending merely on the particular standard of rest which it pleases us to adopt. Whether the apparatus is moving or not, and in what direction, is a question not of objective fact but of arbitrary choice. What we have done is not to explain the unobservability of a natural occurrence ; it is to describe observable natural occurrences in such a way that unobservable ones are not implied.

Necessity for a Generalized Definition of Length. The apparent arbitrariness of the substitution of $l\sqrt{1 - v^2/c^2}$ for l tends to disappear when we reflect on the general nature of physical formulae. We make measurements of various kinds, between which we seek simple relations. Some measurements, like that of length —which is merely the number of times a chosen unit rod is contained in the distance between two points—are so simple as to seem inevitably of fundamental importance. Others, such as that of entropy, are so far from being obviously important that we have to reach an advanced stage in the development of physics before we realize that they are important at all. Nevertheless, when their significance has been recognized we use them freely and they play as prominent a part in our expressions of laws of nature as do the others.

It is a common experience that when we enlarge the scope of the phenomena we consider, quantities that formerly seemed fundamentally important are discovered to be only special forms of more general ones. Let us take an example.

If we allow a given wire to join two points whose electrical potentials differ by E, and measure the current, C, which passes through the wire, we find that the ratio of E to C is constant, whatever the individual values of E and C. This we take to be an important law of nature (Ohm's law), and we write

$$\frac{E}{C} = R \qquad \qquad (14)$$

Within the limits of experiment so far considered, then,

$\dfrac{E}{C}$ is a quantity of fundamental importance, and we give it a name—the *resistance* of the wire.

Let us now, however, extend the scope of our inquiry beyond the mere variation of E and C, by stretching the wire into different shapes. We now find that the simple relation (14) no longer holds, but a more complicated expression assumes importance. If l and σ denote the length and cross-section of the wire, we find the relation

$$\frac{E\sigma}{Cl} = \rho, \qquad . \quad . \quad . \quad . \quad (15)$$

where ρ is constant. We accordingly call ρ the *specific resistance*, and recognize $\dfrac{E\sigma}{Cl}$ as a quantity of greater significance than $\dfrac{E}{C}$, which is merely the special form it assumes when l and σ have certain fixed values.

All this is supposed to take place at a constant temperature. If, now, we still further extend the scope of our inquiry to cover a finite range of temperature, even the specific resistance is seen to have only a limited importance, and the more general expression,

$$\frac{E\sigma}{Cl(1 + \alpha T)} = \rho_0, \qquad . \quad . \quad . \quad (16)$$

gives us a still more fundamental physical magnitude.

We need not carry the illustration further, because the point is probably clear that measurements which are important in a restricted field of research have to be generalized for application to a larger field. The greater part of physics has been concerned with bodies whose relative velocities have been small, and the measurement of length, l, has taken part in many formulae. When the inquiry is extended to large velocities it is therefore quite in keeping with what we might expect that a more complex expression, such as $l\sqrt{1 - v^2/c^2}$, should take its place.

For the precise form of this generalization we have to depend ultimately on observation, but we can go one step further in reasoning and see why it must involve v. In all descriptions of physical processes we must adopt some standard of rest; otherwise we can give no meaning to velocity. The principle of relativity says that there is no natural, absolute standard, so that, apart from considerations of convenience, we can choose any we like, and we shall then mean, by the velocity, v, of a body, the velocity relative to that standard. Let us suppose we have made our choice (e.g. the laboratory) and expressed the physical relations we have discovered in such a way that all velocities are relative thereto. Now choose another standard (e.g. the Sun). Our observations will be precisely the same, otherwise there would be a physical effect arising from a change of motion which leaves all *relative* velocities of bodies unaltered: our laws of nature—i.e. our expressions of physical relations between observations —must therefore be the same. But obviously our expressions of particular quantities will not be the same; for example, whereas in the first case the laboratory bench has velocity zero, in the second case it has a velocity of about $18\frac{1}{2}$ miles a second in a certain direction. Consequently there must be a change in the expressions of other physical quantities, so that when all such quantities are combined together into a law of nature, the modifications shall cancel one another out.

An example drawn from a kind of relativity—relativity of position—which is familiar and universally acknowledged, may make the point clear. We know that, in any problem involving geometrical considerations, we may choose our origin of co-ordinates at any point we like without affecting the solution. Let us, then, make a certain choice, according to which the position of a particle (we consider one co-ordinate only, for simplicity) is x. If, now, we change the origin to a point distant α from the first, the position of that particle is no longer x but $x - \alpha$, and the positions of all the other parts of the system under consideration must be modified accordingly. Our general expression of position must therefore be not x but $x - \alpha$. If we make this substitution for *all* statements of position we shall then be automatically protected against the danger of giving a false significance to a change arising merely from an arbitrary change in our co-ordinate system. For instance, if we wish to find the distance between points x_1 and x_2, we have to write $(x_2 - \alpha) - (x_1 - \alpha)$, and we get the same result whatever value α might have.

Now change the relativity in this example from that of position to that of velocity. Our arbitrary original choice of co-ordinate system now becomes an arbitrary original choice of standard of rest; and the change to a system distant α from the first becomes a change to a system moving with velocity v with respect to the first. The generalization of specification of positions from x to $x - \alpha$ accordingly becomes a generalization of specification of lengths from l to $l\sqrt{1 - v^2/c^2}$.

One point in this analogy may arouse question. In the case of relativity of position, the generalization was that of statements of position, and this seems quite natural. But in the case of relativity of velocity the generalization proposed is not that of statements of velocity but that of statements of length. Why is this?

The answer is that there is a generalization of statements of velocity also, and if we had approached the matter from the point of view of the Fizeau experiment (which will be discussed in terms of relativity in Chapter V) instead of the Michelson-Morley experiment, that would have been the generalization at which we should have arrived, though with much greater difficulty. But velocity, according to the traditional definition of physics, is not a primary quantity : it is measured by the rate of change of position with time, and so involves measurements of both length and time. It follows that a generalization of the specification of velocity must involve a generalization of the specification of length or time or (as, in fact, is the case) both. It is therefore simpler in the long run to deal first with the generalization of the more elementary measurement, and to deduce therefrom that of other measurements.

In this connexion it should be noted that the unit of length is the only unit in physics that is defined independently of any other unit. We choose a particular piece of metal, make two marks on it, and take the distance between them as our unit of length. There is a tendency to substitute for this unit another, which is called the wave-length of monochromatic light, but whichever we choose it is not necessary to be able to measure any other physical quantity in order to measure length. That is not true of any other unit. As we shall see, the conventions of physics are such that we cannot measure time, mass, velocity, force, or any other physical magnitude without being able to measure length at least, and if we can properly generalize the pre-relativity measurement of length, we can then deduce the *consequent* generalizations of all other measurements.

It does not necessarily follow that such generalizations are the only ones to which these other measurements must be subjected, but we find that, in fact, they are so. We shall see that the whole of the requirements of the special theory of relativity are met by the simple substitution of $l\sqrt{1 - v^2/c^2}$ for l at the very beginning of physics. This is not sufficiently realized. It is often supposed, in particular, that an *independent* generalization of the measurement of time is required. As this idea is probably responsible for the widespread illusion that relativity has something mystical to do with the nature of time, and requires that time shall be transformable into space, or some such nonsense, it is important to emphasize this. The special theory of relativity is *completely* contained in the purely physical statement that the fundamental measurement of physics is $l\sqrt{1 - v^2/c^2}$, all other measurements which in classical physics have been defined in terms of l being thereby subject to modification only by the substitution of this more complete expression, their definitions remaining otherwise the same. The remaining chapters of this book will be concerned with the derivation of the proper modifications.

The Question of Terminology. Whenever a particular conception becomes generalized, the question of terminology arises. In the case before us we have to substitute for l the expression $l\sqrt{1 - v^2/c^2}$, where l is what has usually been called the 'length' of an object, measured by a scale at rest with respect to the object, and v is the velocity of the object in the direction of its length, relative to our arbitrary standard of reference.

We are now in a dilemma. Shall we continue to give the name 'length' to l, or shall we transfer it to $l\sqrt{1 - v^2/c^2}$? If we adopt the former alternative, we must say that the length of a body does not change with motion, but pays for its invariance by becoming a quantity of no particular importance. If, on the other hand, we give the name 'length' to the expression $l\sqrt{1 - v^2/c^2}$, the length of a body retains its importance, but we must regard it as changing with velocity. It is the latter meaning of length that custom has decided we shall adopt.

The implication of this choice is often expressed by the statement that a body contracts on moving, but the expression is unfortunate : it suggests that something happens to the body, whereas the 'movement' may be given it merely by our mental change of standard of rest, and we can hardly suppose that the body shrinks on becoming aware of that. We can properly say that the *length* of the body contracts, because the length is not an intrinsic property of the body, but a conception which we associate with the body, and we define it as a function of l and v which decreases in value as v increases. The statement that the *body* contracts on moving is an expression of the Fitzgerald contraction hypothesis, and that, as we have seen, is fundamentally different from the hypothesis of relativity.

The attempt has frequently been made to express this difference by saying that, on the contraction hypothesis, it is the body that changes, whereas relativity requires that ' space ' changes. This well-meaning attempt to point out a real distinction must rank among the major errors of scientific exposition. It at once leads us to think of something outside observable physical facts, which alone are our proper concern, and to imagine a subtle transformation of a metaphysical emptiness endowed for the purpose with inapprehensible qualities which undergo a mysterious transformation. All this is utterly foreign to the matter. We have no need to introduce space into the discussion at all (the four-dimensional ' space-time continuum ' will be considered in Chapter VIII) ; our province is simply that of physical *measurements*, and our object is simply to relate them with one another accurately and consistently. In the field of experiment with which we are now dealing, this is completely achieved by the re-definition of ' length ', and there is no need to think at all of the nature of space, or of anything happening to it.

Transformation of Co-ordinates. Geometrical problems are now universally dealt with by reference to co-ordinate systems, and it is therefore necessary to consider what effect our new definition of length will have on the equations of transformation from one co-ordinate system to another. The transformations in which we are interested are, of course, those which are necessary when we change to a co-ordinate system which is moving with uniform velocity, v, with respect to the original one, and since we can always choose the direction of our axes to

suit our convenience, we shall not lose generality by considering only the case of two Cartesian co-ordinate systems
with parallel axes, of which one—K', $O'(x',y',z')$—is moving with uniform velocity, v, in the positive x-direction
with respect to the other—K, $O(x,y,z)$—which is arbitrarily taken to be at rest (Fig. 2).   We suppose further that
the origins and axes of the systems coincide at some
instant which we will call $t = 0$.

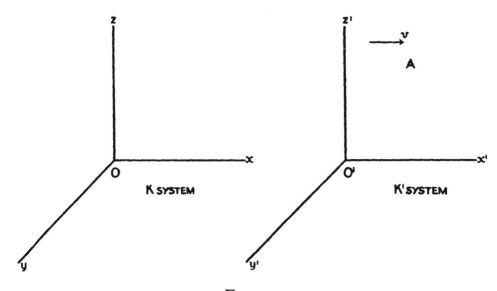

FIG. 2.

Consider a point, A, at some instant, t, when O' has
travelled a distance vt, from O.   Let its co-ordinates be
(x,y,z) in the K system, and (x',y',z') in the K' system.
We want to express (x',y',z') as functions of (x,y,z).   In
terms of the old definition of length we should have the
familiar transformation equations :

$$\left.\begin{aligned} x' &= x - vt \\ y' &= y \\ z' &= z \end{aligned}\right\} \quad . \quad . \quad . \quad . \quad . \quad (17)$$

But these must now be modified.   It must be noted that
x', y' and z' represent positions and not lengths, so we

cannot apply the correction factor directly (see next paragraph). It is introduced in the following way. The magnitudes of the co-ordinates are assigned in terms of a standard length—a standard measuring rod, if we prefer to think in material terms—and if the length of this rod be taken as unity in the K system, it must be taken as $\sqrt{1 - v^2/c^2}$ in the x'-direction of the K' system, since it there has velocity $- v$. Accordingly, the above value of x' must be divided by $\sqrt{1 - v^2/c^2}$ to obtain the number of times the rod is contained in the instantaneous distance x', and we then have

$$\left.\begin{aligned} x' &= \frac{x - vt}{\sqrt{1 - \dfrac{v^2}{c^2}}} \\ y' &= y \\ z' &= z \end{aligned}\right\} \quad \cdots \quad (18)$$

(We shall continually be referring to the K and K' systems, and shall reserve the symbol v for their relative velocity.)

Meaning of the Transformation Equations. The proper interpretation of these equations is a matter of great importance for the understanding of relativity theory. The direct meaning is that if the co-ordinates of a point, A, in any co-ordinate system, K, are (x,y,z), and we then change the co-ordinate system to another, K', which moves with uniform velocity, v, with respect to the first, as above described, then we must take (x',y',z'), as in (18), for the co-ordinates of A in this system in order to ensure that our equations shall not imply the possibility of observing effects which are actually unobservable.

We must notice, however, that the equations hold only for a certain instant of time, viz. t. It is not true to say, for instance, that if K be fixed in a railway station, and K' in a train passing through it, and if (x,y,z) be the co-ordinates of the station clock in K, then the co-ordinates

of the clock in K' will be (x',y',z').   In K' the clock will
not be at a point but spread out along a line, for it will be
moving with respect to the train.   It is only when we
restrict our consideration to a particular instant of time
that we can apply the formulae (18).

It follows that the formulae take no account of whether
any object which may be situated at A, or of which A may
be a part, is moving or at rest in either system.   It is true
that velocity is so defined that we may speak significantly
of the velocity of a body at a point at some instant, but in
the derivation of the transformation formulae we make
no assumption concerning that velocity. A is a ' point
instant ';   it is merely the instantaneous position of a
particle which may have both instantaneous position and
instantaneous velocity, and the formulae hold good for it
whatever the velocity might be.   It is therefore meaningless to say, for instance, that the point to which they refer
is at rest in either system.   That is why, in deriving the
formulae, we could not take what might have seemed the
obvious course and modified the classical formulae (17)
by simply multiplying the right-hand side by $\sqrt{1 - v^2/c^2}$.
Since A's motion might have been anything with respect
to K or K', neither x nor x' could be treated as a ' length '
with which a velocity could have been associated.

It is for this reason that the concept known as the
' observer ' plays such an important, and in some respects
unfortunate, part in relativity literature.   If we had not
adopted the device of describing physical processes in
terms of co-ordinate systems, the observer need not have
appeared at all.   Every space measurement would have
been of the nature of a length, and the value assigned to
that length would have depended on its velocity with
respect to the arbitrary standard of rest chosen.   But now
that we are dealing with co-ordinates, which have no
unique velocity, the v introduces itself as the relative
velocity of the co-ordinate *systems*, and it is sometimes very

convenient to identify a co-ordinate system with an observer who is at rest in that system. We then often say that the observer ' uses ' the system in which he is at rest.

There are, however, in this custom dangers against which we must be on our guard. A particular observer need not necessarily use the system in which he is at rest : any observer may use any system, and, in fact, a single observer frequently changes his system. In certain problems of celestial dynamics, for instance, we usually regard the Sun as being at rest, but we record the positions of the planets according to a system in which the Earth is. at rest. In the former case we may, if we wish, assume that it is not we who are assuming the Sun to be stationary, but an observer on the Sun ; but not only have we no reason to believe there is an observer on the Sun and very good reason to believe there is not, but also the phraseology tends to create the idea that there is something different in the *experience* of relatively moving observers. When we express the ' contraction ' of a moving rod, for instance, by saying that ' *A* observed *B*'s rod to be shortened ', it is difficult not to imagine that there is some visible difference in the appearance of the world, according to whether we take our stand with *A* or with *B*. But it is the fundamental requirement of relativity that there is no such difference : if there were, we could use it to distinguish *A*'s state of motion from *B*'s, and so obtain a physical criterion of motion.

It is much less objectionable to represent co-ordinate systems by ' measuring instruments ' instead of ' observers ', for that introduces no false psychological idea, although it still leaves room for the error that there is some physical difference between relatively moving instruments. We must continually remind ourselves that what the equations imply is that if we change our co-ordinate system from K to K' (that is, in physical instead of mathematical terms, if we change our standard of rest to one moving with velocity v with respect to the original one), the effect, on our description of the world surveyed, of the consequent change in the value we must assign to the length of our measuring rod, is expressed by substituting (x',y',z') for (x,y,z) according to (18).

CHAPTER IV

TIME

THE remainder of our discussion is concerned with the effect, on the measurement of the various physical quantities, of our re-definition of length. We begin with the effect on the measurement of time.

Interdependence of Time- and Space-Measurement. Any procedure for measuring time must involve the choice of some recurring phenomenon, the successive occurrences of which are said to occupy, or to be separated by, equal times. In principle any such phenomenon may be chosen, but since our present purpose is not to re-mould the whole scheme of physics but simply to see how the established scheme must be amended because of the single modification of the definition of length, our first object must be to identify the system of measuring time which physicists actually recognize.

The recurring phenomenon which has been chosen is the rotation of the Earth with respect to the fixed stars : one such rotation is called a *sidereal day*, and all sidereal days are taken as equal intervals of time. By processes which do not concern our inquiry, a fraction of the sidereal day, known as the ' mean solar second ', is adopted as the *unit* of measurement, and successive mean solar seconds are officially regarded as equal intervals of time.

The word ' officially ' is important, for in practice we adopt a time-scale according to which the duration of the mean solar second slowly increases as time goes on. This comes about in the following way.

Taking the standard metre as a constant unit of length, and the mean solar second as a constant unit of time, and assuming that we know of all the forces which may disturb the motion of a body, Newton's first law of motion (' every body perseveres in its state of rest or of moving uniformly in a straight line, except in so far as it is made to change that state by external forces ') becomes a hypothesis to be tested by experiment. We can, in principle, take two bodies, on which no forces act after we have set one in motion relatively to the other ; measure the number of metres covered by the former with respect to the latter in successive seconds ; and see if the numbers are equal. If they are, we can take the law as expressing a fact of observation. If they are not, we can take one of two courses : either we can say that the law is false, or we can change our time-scale to one defined by the moving body itself (i.e. one in which equal times are *defined* as those in which the body moves over equal distances). In the latter case the law automatically becomes true, but we must then say that the mean solar second is not a constant interval of time.

The experiment in this elementary form cannot be made, but we have its equivalent. Laws of mechanics based on Newton's law enable us to calculate the paths of totality of solar eclipses in ancient times. There are records of some of these eclipses, which can be compared with the calculations. Since the Earth's velocity of rotation at the equator is some 16 miles a minute, and the records extend backwards for over 1,000 years, it is clear that a comparison of the calculated and recorded places of visibility of the eclipses affords a very delicate test of the laws. It is found, in fact, that there is a slight discrepancy.*

We now have the choice above mentioned : we can discard the laws or amend our time-scale. The latter is

---

* See Oppolzer, *Canon der Finsternisse.*

the alternative adopted, though without formal acknowledgment. We say that the rotation of the Earth is slowing down at the rate of about $\frac{1}{1000}$th of a second per century, which is just the amount necessary to keep the laws true ; but we have not, in fact, officially discarded the mean solar second as the unit of time. For all practical purposes it is, of course, unnecessary to do so, because the error involved is so extremely minute, but from the point of view of physical theory the change is of the most profound importance. Its effect on our present problem is this : that instead of measuring time independently of other quantities, we make its measurement depend on the measurement of space. Our unit of time is now *defined* as the time taken by a particular kind of moving body to cover a chosen number of units of *length*, and the modification of the definition of length which in the last chapter we saw we had to make, therefore *necessarily* implies a modification of the unit of time.

What the modification is can easily be deduced. Let us choose a standard of rest, and define a unit of time in the way just described. To fix our ideas, suppose the moving body is a ball rolling without friction along a graduated space-scale at rest with respect to our standard, and suppose that t units of time correspond to a movement of the ball over a length l of the scale. Now choose another standard of rest, with respect to which our apparatus is moving with velocity v. Then instead of l we must take $l\sqrt{1 - v^2/c^2}$, and therefore instead of t we must take $t\sqrt{1 - v^2/c^2}$. Hence the transformation equation for time measurement is the same as that for length measurement.

Clocks. A very familiar expression of this result is the statement that the rate of a clock is changed by motion, and by this we are intended to understand that some physical change occurs in the clock. How false this is can be seen, just as we saw the falsity of the corresponding statement for space-measuring rods, by remembering that we can change the velocity of the clock merely by changing our minds.

There is, however, an additional objection to the statement with regard to clocks ; namely, that it is meaningless as well as false, if the phrase may be pardoned. The definition of the unit of length (in a system at rest) is at the same time a definition of the instrument for measuring length—we cannot prescribe an interval between two marks on a bar without prescribing the bar—but the definition of the unit of time involves no definition of the instrument (the ' clock ') for measuring time. Any mechanism—a properly graduated sundial, a water-clock, a pendulum, &c.—which records intervals of time according to which the laws of mechanics are obeyed is a ' clock ', and, apart from convenience, none of them has any claim to be preferred to another. The statement that the rate of a clock is affected in a particular way by motion is therefore meaningless unless it happens by chance that all clocks are affected in the same way. That this is not so may perhaps be sufficiently evident if we think of a moving sundial, but if, through some prejudice, one is not inclined to accept this as a fair case, it may be remarked that a comparison of the behaviour of other mechanisms shows that at least some of them differ in their behaviour when moving similarly.* It follows that the transformation of the unit of time which we have derived is not a statement concerning the readings of clocks : it is, in fact, a statement of the change which we must make in our choice of a unit of time if we change our choice of a standard of rest. What the clock does when we change our minds or when we compare it with another, relatively moving, clock is quite an irrelevant matter.

The Lorentz Transformation for Time. We have

* See *Nature*, **144**, p. 888 (1939).

seen that the Lorentz transformation formulae apply only to point-instants, and not to places irrespective of times. The instant t to which they apply is therefore of the nature of a fourth co-ordinate, and there will be a transformation formula relating t to t', corresponding to the formulae relating (x,y,z) to (x',y',z').

We can deduce this formula directly from the modification already found for the unit of time, combined with the known formulae for the space co-ordinates, but the work may be shortened by making use of the fact, revealed in the Michelson-Morley experiment, that the velocity of light must have the same value in all co-ordinate systems ; otherwise we should be able to determine our motion through the ether by finding its maximum and minimum values. Referring to our systems, K and K', then, let us suppose that at the instant $t = 0$, when the axes momentarily coincide, a beam of light is emitted from the origin in the x-direction, and that this beam reaches the point (x,y,z), or (x',y',z'), at the instant t, or t'. We must then have

$$c^2 = \frac{x^2}{t^2} = \frac{x'^2}{t'^2};$$

i.e.
$$c^2 t^2 - x^2 = c^2 t'^2 - x'^2. \qquad . \quad . \quad . \quad . \quad (19)$$

Substituting for x' in terms of x from (18), and solving for t', we obtain

$$t' = \frac{t - \frac{v}{c^2}x}{\sqrt{1 - \frac{v^2}{c^2}}} \qquad . \quad . \quad . \quad . \quad (20)$$

Combining this with the equations for the space co-ordinates, we have the following transformation formulae for point-instants :—

$$\left.\begin{aligned} x' &= \frac{x - vt}{\sqrt{1 - \dfrac{v^2}{c^2}}} \\[2em] y' &= y \\ z' &= z \\[1em] t' &= \frac{t - \dfrac{v}{c^2}x}{\sqrt{1 - \dfrac{v^2}{c^2}}} \end{aligned}\right\} \quad \cdots \cdots \quad (21)$$

These are the famous Lorentz transformation formulae, derived before the advent of the special theory of relativity, but only to be fully understood in the light of that theory.

Significance of the Transformation. It is important to understand exactly how time comes to be associated with space in these equations. The association does not at first appear necessary. We have seen that the Michelson-Morley experiment is quite independent of time measurement, and, further, that the whole of the requirements of the special theory of relativity are satisfied by the re-definition of length. It is true that we have explained in detail in this chapter why a re-definition of length must involve a re-definition of the unit of time measurement, but since the Michelson-Morley experiment has nothing to do with time measurement, and the transformation formulae have been derived from a modification of length which could have been deduced from that experiment alone, the problem still remains why time and space are so intimately associated in equations (21). It will be noticed that the expression for x' involves t, and that for t' involves x, so that the two kinds of measurement appear inextricably entangled.

The explanation is that in dealing with relative *velocities*, we are dealing with something which implicitly involves time. This is not demanded by the nature of things, but arises from the way in which we have chosen to measure velocity—namely, as the rate of change of position with time ; but that does not alter the fact that it is so. Consequently, a principle which says there is no meaning in absolute *velocity* is a principle which says something about space and time. In the Michelson-Morley experiment, for instance, we take v to be constant, but, the method of measuring space being prescribed, a given velocity is constant only if a certain time-scale is assumed. If it is constant when the mean solar second is the unit, then it is not constant when the actual unit of physics is chosen, and *vice versa*. (In this example, of course, the difference would be too small for detection, but the principle is sufficiently illustrated.) Our fundamental modification of length, therefore, by introducing v, automatically commits us to the particular time-scale by which v is constant. When we come to the transformation formulae, this appears in the term vt in (17) ; that is the (classical) distance between the axes of co-ordinates only when t is the time interval measured according to the time-scale by which v is constant.

The intimate association of time with space which relativity is usually credited with having established is therefore nothing but a consequence of the voluntary acts of the pioneers of physics. A material rod having been chosen as the measure of space, a time-scale might have been chosen in terms, say, of the periodic fluctuations in the light of a variable star, and velocity might have been measured by the static deflections of the pointer of a speedometer. In that case there would have been no necessary connexion between space, time and velocity measurements, and any relation found between them would have been an empirical law of nature. But this has not

been done : time has been measured in terms of the
measure of space, and velocity in terms of the measures of
time and space, and the consequence is that the three sets
of measures are connected by definition and not by nature.
Moreover, relativity has not united them more closely :
it has simply made a slight modification of the former
expression of the association. In the classical equations
(17), the expression for x' involves t no less than do the
relativity equations (18). All talk of the nature of space
and time is therefore quite irrelevant to relativity, which
is as purely physical as the kinematics of former days.

Applications of the Formulae. If we consider
two neighbouring point-instants, we may express the
transformation formulae for the interval between them by
differentiating (21), whence we obtain

$$
\left.
\begin{aligned}
dx' &= \frac{dx - vdt}{\sqrt{1 - \dfrac{v^2}{c^2}}} \\
dy' &= dy \\
dz' &= dz \\
dt' &= \frac{dt - \dfrac{v}{c^2}dx}{\sqrt{1 - \dfrac{v^2}{c^2}}}
\end{aligned}
\right\}
\qquad . \quad . \quad . \quad . \quad (22)
$$

Let us apply these equations to one or two simple cases.

First, consider what happens at a point fixed in the
K system when we change to the K' system. In this case
we put $dx = 0$, and obtain

$$
dx' = - \frac{vdt}{\sqrt{1 - \dfrac{v^2}{c^2}}} \qquad . \quad . \quad . \quad . \quad (23)
$$

a quantity which increases numerically with dt, and therefore with dt', from (24). This is the common-sense fact

that a point fixed in the laboratory, say, becomes a finite part of the Earth's orbit when we change the standard of rest to the Sun, its length depending on the interval of time, dt', which we care to consider. A less familiar result comes from the last of equations (22), viz.

$$dt' = \frac{dt}{\sqrt{1 - \dfrac{v^2}{c^2}}} \qquad . \quad . \quad . \quad . \quad (24)$$

This expresses the modification of time intervals which we have already deduced directly from the fundamental re-definition of length. For suppose the two point-instants are marked by successive ticks of a clock stationary in the K system (the laboratory) and therefore moving in the K' system. Then since, as we have seen, the unit of time in the K' system is $\sqrt{1 - v^2/c^2}$ times that in the K system, the number of units between two particular events must be $\dfrac{1}{\sqrt{1 - \dfrac{v^2}{c^2}}}$ times greater in the K' system than in the K system, and this is equivalent to (24).

Now consider an *instant*, instead of a point, in the K system. We must then put $dt = 0$, and we obtain

$$\left. \begin{aligned} dx' &= \frac{dx}{\sqrt{1 - \dfrac{v^2}{c^2}}} \\[2em] dt' &= -\frac{\dfrac{v}{c^2}dx}{\sqrt{1 - \dfrac{v^2}{c^2}}} \end{aligned} \right\} \quad . \quad . \quad . \quad . \quad (25)$$

The second of these equations expresses one of the most unexpected requirements of relativity—namely, that two events simultaneous in one system of co-ordinates are not

simultaneous in a relatively moving system. The surprising character of this result is somewhat mitigated when
we notice that it applies only to events separated in space
in both systems, for if dx or dx' vanishes as well as dt,
then equations (25) show that dt' vanishes also. Hence
events which occur at the same time and place in one
system do so in the other also. But if dx is finite, then
simultaneity in one system is not simultaneity in the other.

Our first impression, that this is contrary to experience,
can be seen to be erroneous. All our experience happens
at a single point—our brain—so that any two observations
we make are separated by a zero space interval in the
system in which we are at rest. If, then, in that system,
$dt = 0$ also, the observations are simultaneous in *every*
system. But when we are dealing with events at *different*
places—e.g. the appearance of a new star and the sound
of a gun—the determination of the times of their occurrence involves *calculation*; and when we change our
co-ordinate system, the formulae on which the calculations
are based are changed. Hence while in one system the
events may turn out to have been simultaneous, they will
not necessarily be found so in the other. That is all that
is meant by saying that simultaneity is relative and not
absolute. It has nothing to do with experience, but only
with calculations.

The first of equations (25) is the equivalent, in terms
of co-ordinates, of our fundamental substitution of
$l' = l\sqrt{1 - v^2/c^2}$ for l, but we cannot immediately identify
dx and dx' with l and l' for the following reason. l is the
length of a body as measured by a rod with respect to
which it is at rest, but l' is the length as measured by a
rod with respect to which it is moving. To make the
latter measurement, therefore, we must stipulate that the
two ends of the measuring rod shall be read at the same
instant; otherwise, since the measuring rod passes con-
tinuously along the body, we can get whatever value we

please for l' by choosing our moments for taking the readings. If, then, dx represents l, dx' given by (25) does not represent l' because it is the distance between the ends of the body when $dt = 0$ and not when $dt' = 0$.

To deduce the 'contraction' of a moving rod from equations (22), therefore, we must re-express them, to obtain dx and dt in terms of dx' and dt'. This we can do either by elementary algebra, or immediately by interchanging dashed and undashed symbols and changing the sign of v. We then obtain

$$\left. \begin{aligned} dx &= \frac{dx' + vdt'}{\sqrt{1 - \dfrac{v^2}{c^2}}} \\ dy &= dy' \\ dz &= dz' \\ dt &= \frac{dt' + \dfrac{v}{c^2}dx'}{\sqrt{1 - \dfrac{v^2}{c^2}}} \end{aligned} \right\} \qquad \dots \quad (26)$$

and from the first of these equations, by putting $dt' = 0$, we obtain

$$dx' = dx\sqrt{1 - \frac{v^2}{c^2}}, \quad \dots \quad (27)$$

which is the required result.

The Doppler Effect. A particularly important application of the transformation formulae is that which expresses the Doppler effect ; i.e. the difference in frequency of the light from two sources which are approaching or receding from one another but are otherwise identical.

The familiar expression for a train of light waves in terms of a co-ordinate system is

$$y = A \sin\left\{\frac{2\pi}{\tau}\left(t - \frac{x}{c}\right) + \alpha\right\}, \quad \dots \quad (28)$$

where x and t are the space and time co-ordinates, respectively, in a system in which the source of light is at rest, and A, α and τ represent respectively the amplitude, phase and period of the vibration.   It is not necessary, of course, that we should think of this equation as representing a wave motion in the space between the source of light and the observer, although it is frequently convenient to do so The facts of observation are all equally well expressed if we regard the various quantities as representing something characterising the source of light itself.

Now suppose we change the co-ordinate system to one moving with uniform velocity v with respect to the former : then in this system the source of light moves with velocity $-v$, and we must substitute x', t' for x, t, according to equations (21).   We then obtain, after a little reduction,

$$y = A \sin \left\{ \frac{2\pi}{\tau'}\left(t' - \frac{x'}{c}\right) + \alpha \right\}, \qquad . \quad . \quad . \quad (29)$$

where
$$\tau' = \tau \sqrt{\frac{1 + \dfrac{v}{c}}{1 - \dfrac{v}{c}}} . \quad . \quad . \quad . \quad . \quad (30)$$

This means that we may continue to associate light with a periodic vibration in the source if we are prepared to allow that the period of the vibration changes with motion in accordance with (30).

Now if we had simply a single source of light, whose velocity with respect to our spectroscope was constant, this would express no observational possibilities. We could assign to that source whatever velocity we pleased, and express the period of the light in terms of it.   But suppose we have two similar sources of light, one approaching the other with velocity v.   Then, in order to be able to maintain the association of light with a periodic motion in its source, it would be necessary for us to observe the equivalent of a difference of period between the sources. v in that case would be the relative velocity of one source with respect to the other, and not the absolute velocity of either. It is to be noticed, too, that, unlike previous modifications arising from the relativity requirement, equation (30) contains v to the first power, and the change of period is therefore in opposite senses when the sources of light are approaching and receding from one another, respectively.

This difference of period is the well-known Doppler effect, and its existence, from the point of view of relativity, is not so much a confirmation of the theory as a justification for associating light with a periodic motion expressible in terms of a co-ordinate system by (28). The difference of period here is not one which we can bring about merely by changing our minds. We can regard either source as being at rest or moving with any arbitrary velocity, but we cannot, by any such exercise of our liberty, remove the *relative* velocity of one source with respect to the other. We are thus dealing with a genuine physical effect.

But that effect is not a change occurring in either source of light. We cannot say that the frequency is either increased or decreased by motion. For consider three similar sources of light, A, B and C, lying along a straight line and at first relatively at rest. An observer, fixed with respect to B, finds that their periods are equal. Now let him move towards A. He then observes that the period of A is less, and that of C greater, than that of B. Since A and C have remained undisturbed and B has been set in motion, whatever change has occurred to a source of light must be located in B. But that is impossible, for any change of the period of B alone would make it either greater or less than the common period of A and C, whereas it becomes greater than one and less than the other. We must therefore either suppose that a disturbance of B

magically brings about different changes in the undisturbed A and C, or else associate the phenomenon with relative motion alone, and not with intrinsic properties of the light sources.   The latter is the alternative in keeping with modern physical ideas.

The Doppler effect is not a discovery arising out of the theory of relativity.   It was known long before, and seen to be necessarily involved in the association of light with a periodic motion in the source.   The formula arrived at, however, was not (30) but one which, for all but the largest velocities, is almost identical with it, viz.—

$$\lambda' = \lambda\left(1 + \frac{v}{c}\right).$$

Here we have substituted wave-lengths for the corresponding periods, to which they are proportional.

It is instructive to notice that our fundamental redefinition of length immediately converts this into (30). The length λ' is associated with a source conceived to be moving with velocity $-v$, so that we must substitute $\lambda'\sqrt{1 - v^2/c^2}$ for it, while λ, associated with a source conceived to be at rest, remains unchanged.   We then obtain immediately

$$\lambda' = \lambda\sqrt{\frac{1 + \dfrac{v}{c}}{1 - \dfrac{v}{c}}},$$

which is equivalent to (30).

The Aberration of Light.   Another outstanding phenomenon expressible directly by means of the transformation formulae (21) is the aberration of light.

Let the K system be stationary with respect to the Sun, and the K' system stationary with respect to the Earth: the x and x' axes are accordingly in the direction of the Earth's orbital motion (the much slower rotational motion being neglected for simplicity). Choose the moment $t = 0$, when the systems of axes are momentarily coincident, and consider the observation of a star at a point (x,y,z) at that moment. This is the observation of an event occurring at an instant $t = -r/c$, where $r \equiv \sqrt{x^2 + y^2 + z^2}$ is the distance of the star. These co-ordinates, of course, refer to the K system, and if θ be the angle between the direction of motion of the Earth (the x-axis) and the direction of the star, we have

$$\cos \theta = \frac{x}{r}. \qquad \ldots \quad (31)$$

In the K' system, if θ' be the corresponding angle, we have similarly

$$\cos \theta' = \frac{x'}{r'}, \qquad \ldots \quad (32)$$

where

$$\left. \begin{array}{l} x' = \dfrac{x + v\dfrac{r}{c}}{\sqrt{1 - \dfrac{v^2}{c^2}}} \; ; \quad y' = y \; ; \quad z' = z \; ; \\[4mm] r' \equiv \sqrt{x'^2 + y'^2 + z'^2} = \sqrt{x'^2 + r^2 - x^2} \; ; \end{array} \right\} \quad (33)$$

whence we easily obtain

$$\cos \theta' = \frac{\cos \theta + \dfrac{v}{c}}{1 + \dfrac{v}{c} \cos \theta}. \qquad \ldots \quad (34)$$

The direction in which the star appears is therefore different from what it would have been if the Earth had been stationary in its orbit. This is not the place to show that (34) expresses the well-known apparent annual movement of the star along the aberrational ellipse, but the student may easily verify that it does so, if he remembers that the ' displacement ', $\theta - \theta'$, lies in the plane containing the star and the direction of the Earth's orbital motion, and that that direction changes from moment to moment.

This treatment of aberration is not that usually given : it is customary to consider the matter in terms of the velocity of light, whereas the foregoing discussion involves only the positions of the star in the two co-ordinate systems. The two approaches lead, of course, to the same result, and the more usual one is given on page 58, but it seems desirable to emphasize the fact that the formula (34) can be obtained without considering even the existence, let alone the physical nature, of light, because it is still frequently asserted that the phenomenon of aberration necessitates the assumption, or even proves the existence, of an ether. The assertion is groundless, for even the discussion in terms of velocities does not imply that light is a wave motion, but only that it is something moving with velocity c. In the treatment given here, however, no use has been made of the conception of light at all. All that it is necessary to postulate is that if an instantaneous visual observation is said to be that of an event occurring at a point (x,y,z) in an arbitrary co-ordinate system, the event must be assigned to a time $\dfrac{\sqrt{x^2 + y^2 + z^2}}{c}$ before the time of observation, where c is the constant occurring in the definition of length, namely $l\sqrt{1 - v^2/c^2}$. We can, if we wish, clothe this logical skeleton with the flesh and blood of a hypothesis concerning the mechanism by which the postulated distant event becomes the immediate experience of the observation, and the hypothesis of waves travelling in an ether with velocity c is often an extremely convenient hypothesis of this kind, though there are occasions on which it is inappropriate and a corpuscular hypothesis serves the purpose better. We err seriously, however, if we regard any such hypothesis as being established by the fact that it is sometimes convenient to assume it.

CHAPTER V

VELOCITY AND ACCELERATION

VELOCITY is defined as dx/dt; i.e. as the rate of change of position with time. Since the theory of relativity requires a modification of our measurements of position and time which becomes effective when we change our system of reference to one moving relatively to the original one, we may expect a modification of our measurements of velocity which also becomes effective in such circumstances.

This, of course, is very familiar, because, apart from relativity altogether, we know of such a modification. The classical equations (17), supplemented by the time equation, $t' = t$, show that dx'/dt' is not equal to dx/dt when v is not zero. They require, in fact, that if the velocity is measured by V in the K system, it will be measured by $V' = V - v$ in the K' system. What is new in relativity is not the fact that V' is not equal to V, but the particular relation between V' and V.

Transformation Formulae for Velocity. We may obtain this relation at once from (22). If V_x, V_y, V_z, are the components of V, and V_x', V_y', V_z' those of V', we have—

$$V_x' \equiv \frac{dx'}{dt'} = \frac{dx - v\,dt}{dt - \frac{v}{c^2}dx} = \frac{V_x - v}{1 - \frac{vV_x}{c^2}},$$

$$V_y' \equiv \frac{dy'}{dt'} = \frac{dy\sqrt{1 - \frac{v^2}{c^2}}}{dt - \frac{v}{c^2}dx} = \frac{V_y\sqrt{1 - \frac{v^2}{c^2}}}{1 - \frac{vV_x}{c^2}},$$

$$V_z' \equiv \frac{dz'}{dt'} = \frac{dz\sqrt{1 - \frac{v^2}{c^2}}}{dt - \frac{v}{c^2}dx} = \frac{V_z\sqrt{1 - \frac{v^2}{c^2}}}{1 - \frac{vV_x}{c^2}}, \qquad (35)$$

whence

$$V'^2 \equiv V_x'^2 + V_y'^2 + V_z'^2$$

$$= \frac{(V^2 - V_x^2)\left(1 - \frac{v^2}{c^2}\right) + (v - V_x)^2}{\left(1 - \frac{vV_x}{c^2}\right)^2}$$

One interesting point should be noted. Consider two velocities, V_1 and V_2, which in the K system are equal in magnitude but different in direction. These velocities in the K' system will differ in magnitude as well as direction, since V_x will not be the same for both. The preferential occurrence of V_x over V_y and V_z in the equations arises, of course, from the fact that the relative motion of the co-ordinate systems is in the x-direction.

As a special case we may note that if

$$(V_x, V_y, V_z) = (0,0,0),$$

we have $(V_x', V_y', V_z') = (-v, 0, 0)$. This is simply the relation already noted, that if K' moves with velocity v with respect to K, K moves with velocity $-v$ with respect to K'. As a consequence, we may express (V_x, V_y, V_z) in terms of (V_x', V_y', V_z') by simply interchanging the dashed and undashed symbols in (35) and changing the sign of v—a fact that may be verified by algebraic transformation.

We may express (35) in a special form by choosing the x-axis in the direction of motion of the body considered, and confining ourselves, as before, to co-ordinate systems whose motion is in that direction also. We then have $V_x = V$ and $V_x' = V'$, while $V_y = V_z = V_y' = V_z' = 0$, so that

$$V' = \frac{V - v}{1 - \dfrac{vV'}{c^2}}$$

or

$$V = \frac{V' + v}{1 + \dfrac{vV'}{c^2}}$$

$$\quad \cdots \cdots \quad (36)$$

Meaning of the Transformation. To see the meaning of this, suppose the K system is fixed in a railway station and the K' system is fixed in a train moving through the station. Then v is the velocity of the train. Now suppose a passenger rolls a ball along the corridor of the train and measures its velocity with respect to the train. The value he obtains will be relative to the K' system and will be V'. If we suppose that an observer on the station can see the ball through the train and measure its velocity with respect to the station, the value he will obtain will be V. Equation (36) requires that if afterwards they compare notes, they will find their results related by the formula given.

Pre-relativity physics required that V should be the simple sum of v and V'; i.e. that if a body moved with respect to a certain standard with a velocity v, and a projectile was sent out from that body at velocity V' with respect to it, then the velocity of the projectile with respect to the original standard would be $v + V'$. The relativity formula, however, requires a value less than this by an amount which is extremely small for ordinary terrestrial velocities, but which becomes important for velocities comparable with that of light. In particular it will be noticed that so long as v and V' are not greater than c, V also cannot be greater than c. For example, if the velocity of the train is c, and the velocity of the ball with respect to the train is c, the velocity of the ball with respect to the station is also c, and not $2c$ as the simple addition formula would require.

There are two things to be said about this. In the first place, it does not imply that velocities greater than c are not possible : it says simply that we cannot reach such velocities by the composition of any number of velocities which themselves are not greater than c. To suppose that this sets a limit to any single velocity is equivalent to being convinced that Achilles can never overtake the tortoise. The formulae relate essentially to the composition of velocities in different systems of reference : they are not concerned with what can happen in any one system. We shall, in fact (p. 69), see a reason why the velocity of a material body cannot exceed c, but that is quite a different matter.

The second point is that, as we have so often found occasion to remark, this limitation on the magnitude of the velocity obtained by composition has nothing to do with the nature of the universe, but arises from the way in which we have decided to measure velocity. This can be seen most strikingly, perhaps, by such an example as the following. Suppose that velocity had been defined in terms of the Doppler effect instead of as dx/dt. Equation (30), expressed in terms of wave-length λ, gives us

$$\frac{d\lambda}{\lambda} = \frac{v}{c} + \text{higher powers.} \qquad . \quad . \quad . \quad (37)$$

Let us choose a new measure of velocity in which equal increments correspond to equal values of $d\lambda$ for a given value of λ. If V denotes velocity on this scale, we have, by definition,

$$V = \frac{c}{\lambda}d\lambda.$$

Now $d\lambda$ may increase up to infinity, but it cannot decrease below $-\lambda$, for the wave-length of the light cannot become negative. Hence a body may recede with infinite velocity, but cannot approach with a velocity greater than c. Notice that V and v are identical for velocities so small that higher powers of v/c can be neglected, i.e. for most naturally occurring velocities. Our change of definition is thus not a fantastic one, and the example shows that whether or not there is a limit to possible velocities is a characteristic of our scale of measurement, and not of physical possibilities.

Invariance of the Velocity of Light. We see from (36) that if $V' = c$, then $V = c$, whatever v may be : that is to say, the magnitude of the velocity of light is the same in all co-ordinate systems. This we have already seen to be so ; otherwise we could determine our velocity through the ether from the change in that magnitude with motion of the observer. For this reason the quantity c is said to be *invariant*. Again, this is not a physical property of light. To say that c is invariant is not to say that the velocity of light is invariable : we know, in fact, that it is different in air and in water, for example. Mathematically, the word ' invariant ' in this connexion means simply ' unaltered by a Lorentz transformation '. So far as physical phenomena are concerned, the velocity of light may be varied in a number of ways, but we cannot obtain a different value for the velocity *in vacuo* by changing our state of motion when making the measurements. Even this limitation does not apply to the velocity of light in matter, as we know from the Fizeau experiment, which is considered in the next paragraph.

There is another extremely important point to be noticed.

Strictly speaking, it is inaccurate to say even that the velocity of light *in vacuo* is invariant.   Velocity is a vector quantity, having direction as well as magnitude, and it is only the magnitude of the velocity of light *in vacuo* that is invariant to a Lorentz transformation.   This can be seen most directly, perhaps, by considering a beam of light parallel to the y-axis of the K system, so that $V_x = V_z = 0$, and $V_y = c$.   We have then, from (35), $V'_x = -v$; $V'_y = \sqrt{c^2 - v^2}$; $V'_z = 0$.   The motion in the K' system is therefore inclined to the y' axis, and only the total velocity, c, is invariant, not the components of that velocity.

The most important application of this fact is to the aberration of light, which we have already discussed in terms of the instantaneous position of the light-source and which we now discuss in terms of the light proceeding from that source.   Choosing the K and K' systems as on page 50, and adopting the same notation, we consider the velocity of the light in the two systems.   The total velocity is, of course, the same, viz. $-c$ (minus, since velocities of approach are negative).   We have therefore

$$V_x = -c \cos \theta ;$$
and
$$V_x' = -c \cos \theta',$$

whence, using (35), we have at once

$$-c \cos \theta' = \frac{-c \cos \theta - v}{1 + \dfrac{v \cos \theta}{c}},$$

i.e.
$$\cos \theta' = \frac{\cos \theta + \dfrac{v}{c}}{1 + \dfrac{v}{c} \cos \theta}, \quad . \quad . \quad . \quad (34)$$

which is identical with (34).

The Fizeau Experiment.   The result of the Fizeau experiment is explained immediately by an application of formula (36).   The fact revealed by the experiment was that, with respect to the laboratory, the velocity of light in water moving with velocity $\pm u$ was increased by $\pm u(1 - 1/\mu^2)$, and the fact that μ was independent of u showed that the velocity of light with respect to the water was c/μ, whatever the value of u.

Let the K system be fixed in the laboratory and the K' system in the water; then, in (36), $v = u$ (taking the positive sign, for example), and V' (the velocity of light with respect to water) $= c/\mu$. We therefore have for V, the velocity of light with respect to the laboratory—

$$V = \frac{\frac{c}{\mu} + u}{1 + \frac{cu}{\mu c^2}} = \frac{c}{\mu}\left(1 + \frac{u\mu}{c}\right)\left(1 + \frac{u}{\mu c}\right)^{-1} \\ = \frac{c}{\mu} + u\left(1 - \frac{1}{\mu^2}\right) + \ldots \tag{38}$$

The result obtained by Fizeau thus follows immediately from the relativity formula.

Transformation of $\sqrt{1 - V^2/c^2}$. This is a suitable place in which to introduce an algebraic deduction which is of no special importance in itself but will be needed in later calculations. It begins with the transformation of $\sqrt{1 - V^2/c^2}$ in the special case assumed in (36), in which V and v are in the same direction. From that equation we have

$$1 - \frac{V^2}{c^2} = 1 - \frac{1}{c^2}\left(\frac{V' + v}{1 + \frac{vV'}{c^2}}\right)^2,$$

whence, after reduction, we obtain

$$\sqrt{1 - \frac{V^2}{c^2}} = \frac{\sqrt{\left(1 - \frac{V'^2}{c^2}\right)\left(1 - \frac{v^2}{c^2}\right)}}{1 + \frac{vV'}{c^2}};$$

i.e.
$$1 + \frac{vV'}{c^2} = \sqrt{\frac{\left(1 - \frac{V'^2}{c^2}\right)\left(1 - \frac{v^2}{c^2}\right)}{1 - \frac{V^2}{c^2}}}.$$

Now apply this equation to two velocities which, in the K' system, are equal and opposite, so that $V_2' = -V_1'$. We thus obtain

$$1 + \frac{vV_1'}{c^2} = \sqrt{\frac{\left(1 - \frac{V_1'^2}{c^2}\right)\left(1 - \frac{v^2}{c^2}\right)}{1 - \frac{V_1^2}{c^2}}}$$

and
$$1 - \frac{vV_1'}{c^2} = \sqrt{\frac{\left(1 - \frac{V_1'^2}{c^2}\right)\left(1 - \frac{v^2}{c^2}\right)}{1 - \frac{V_2^2}{c^2}}} ;$$

whence
$$\frac{1 + \frac{vV_1'}{c^2}}{1 - \frac{vV_1'}{c^2}} = \sqrt{\frac{1 - \frac{V_2^2}{c^2}}{1 - \frac{V_1^2}{c^2}}} \quad \cdots \cdots \quad (39)$$

It should be noted that since this result is derived from (36) instead of (35), it applies only when the velocities V and v are both in the x-direction. The more general case can easily be worked out, and is left as an exercise for the student.

Transformation Formulae for Acceleration. According to pre-relativity physics, acceleration is invariant with respect to co-ordinate systems in uniform relative motion. This follows immediately from (17), which gives $(\ddot{x}',\ddot{y}',\ddot{z}') = (\ddot{x},\ddot{y},\ddot{z})$. The corresponding relativity expressions are obtained as follows. We have

$$\dot{V}_x' \equiv \frac{dV_x'}{dt'} = \frac{dV_x'}{dt}\cdot\frac{dt}{dt'},$$

with similar equations for $\dot{V}_y'$ and $\dot{V}_z'$. The first factor is evaluated by differentiating (35), and the second directly from (26). We thus find

$$
\left.
\begin{aligned}
\dot{V}_x' &= \frac{\left(1 - \dfrac{v^2}{c^2}\right)^{\frac{3}{2}}}{\left(1 - \dfrac{V_x v}{c^2}\right)^3} \cdot \dot{V}_x \\[2em]
\dot{V}_y' &= \frac{1 - \dfrac{v^2}{c^2}}{\left(1 - \dfrac{V_x v}{c^2}\right)^2} \left\{ \dot{V}_y + \frac{\dfrac{V_y v}{c^2}}{1 - \dfrac{V_x v}{c^2}} \cdot \dot{V}_x \right\} \\[2em]
\dot{V}_z' &= \frac{1 - \dfrac{v^2}{c^2}}{\left(1 - \dfrac{V_x v}{c^2}\right)^2} \left\{ \dot{V}_z + \frac{\dfrac{V_z v}{c^2}}{1 - \dfrac{V_x v}{c^2}} \cdot \dot{V}_x \right\}
\end{aligned}
\right\} \quad . \quad (40)
$$

The corresponding equations for $\dot{V}_{x,y,z}$ in terms of $\dot{V}'_{x,y,z}$ are, of course, obtained by interchanging dashed and undashed symbols and changing the sign of v.

One interesting fact which follows from these equations is that, since the components of acceleration in the K' system involve the components of velocity as well as those of acceleration in the K system, an acceleration which is constant in one system is not, in general, constant in the other. We must not, however, draw from this fact hasty conclusions concerning force, based on our custom of measuring force by acceleration. We shall see in the next chapter that other considerations also have to be taken into account. Equations (35) show that velocity does not share this anomalous character: a velocity whose components are constant in one system is constant in all others moving with uniform relative velocity.

CHAPTER VI

MASS, ENERGY AND FORCE

The Definition of Mass. Mass is often defined as the quantity of matter in a body, and it is measured by weighing the body. Both definition and fundamental method of measurement, however, are wrongly so stated. How inaccurate the definition is will be seen shortly: measurement by weighing, on the other hand, does not give inaccurate results but is permissible only by virtue of what, from the point of view of the special theory of relativity, is the happy accident that the 'inertial' mass of a body is equal to its 'gravitational' mass. When we are considering a fundamental modification of the measurement of mass we must take into account only the fundamental definition.

Mass is ultimately defined in terms of Newton's second and third laws of motion. The most convenient way of applying the laws to the measurement of mass is to consider the impact of two elastic bodies. By the second law, the force acting on each body is measured by the change of motion which it produces, the motion being measured by the product of the mass and the velocity of the body. By the third law, the total force acting on the two colliding bodies is zero. Hence the total change of motion is zero, and we have the law of 'conservation of momentum' for this special case.

For convenience, consider a head-on collision. Let the (at present unknown) masses of the two bodies be m_1 and m_2, and let their velocities (supposed constant) before collision be V_1 and V_2, respectively, with reference to some arbitrarily chosen co-ordinate system. We are not yet at liberty to assume that the masses are independent of the velocities, so let us suppose that after the collision they are respectively $\bar{m}_1$ and $\bar{m}_2$, and that the velocities are $\bar{V}_1$ and $\bar{V}_2$. The law of conservation of momentum then gives us

$$m_1 V_1 + m_2 V_2 = \bar{m}_1 \bar{V}_1 + \bar{m}_2 \bar{V}_2 . \quad . \quad . \quad (41)$$

Now we can measure all the velocities, but not the masses, for at present we know nothing of them except that they must satisfy this equation, which is a single relation between four unknown quantities. We therefore make an additional postulate ; namely, that the total mass, as well as the total momentum, is conserved in the collision. We then have

$$m_1 + m_2 = \bar{m}_1 + \bar{m}_2 \quad . \quad . \quad . \quad . \quad (42)$$

These two equations are sufficient to determine the measurement of mass. For, consider the instant of collision, when both bodies have momentarily the same velocity, V, say. We then have, from (41) and (42),

$$m_1 V_1 + m_2 V_2 = (\bar{m}_1 + \bar{m}_2)V = (m_1 + m_2)V, \quad . \quad (43)$$

whence

$$\frac{m_1}{m_2} = \frac{V - V_2}{V_1 - V} . \quad . \quad . \quad . \quad . \quad (44)$$

We thus obtain the ratio of the initial masses in terms of observable quantities, and by choosing one of them—say m_2—as an arbitrary unit, we can measure the mass of any other body.

All this is traditional physics. It should be noticed that it contains no requirement that m_1 and m_2 shall be equal respectively to $\bar{m}_1$ and $\bar{m}_2$: it is the total mass only that, by definition, is conserved in the collision. It is true that experiments at velocities small compared with c are consistent with the assumption that the individual masses are invariable, and these, supplemented by the experiments on the weights of bodies which had led to the idea of the ' indestructibility ' of matter, gave rise to the belief that mass was not merely a ' primary quality ' of a body, but *the* primary quality—the ' quantity of matter ' in the body. In view of the results we are about to obtain, it is important to remember that this belief depends entirely on experiments conducted within a limited range of conditions : it cannot be deduced from the definition of mass, and is liable to be destroyed either by more comprehensive experiments or by logical deduction from the definition of mass and other established facts.

Mass and Velocity. If we accept the special principle of relativity as an established fact, then it is possible to show that the above definition requires that the mass of a body shall not be independent of its velocity. To see this, we consider the collision experiment just described, in terms of two systems of co-ordinates moving with relative velocity v. We know the transformation equations for the velocities, and we can therefore determine what the transformation equations for mass must be in order to satisfy the definition of mass contained in (44). We shall find that in order that such transformation equations shall be possible at all, the quantities m_1 and m_2 must themselves be functions of the velocities, V_1 and V_2, respectively.

The case of a general collision has been worked out,[*] but, on account of its length, we take only a very special case, which will, nevertheless, yield the general result. Consider a head-on collision between two exactly similar, perfectly elastic bodies. Their similarity ensures that, in the same circumstances, their masses shall be the same, but if we suppose that in the K system their velocities before collision are different,—say V_1 and V_2—the circumstances are not the same, and we must give them possibly different

[*] See Lewis and Tolman, *Phil. Mag.*, **18**, p. 510 (1909).

masses, viz. m_1 and m_2. We then have, by the postulate of conservation of mass,

$$m_1 + m_2 = M, \quad . \quad . \quad . \quad . \quad . \quad (45)$$

where M is constant before, during and after the collision. At the moment of collision the velocities will be instantaneously the same—V, say; and we have therefore, by the postulate of conservation of momentum,

$$m_1 V_1 + m_2 V_2 = MV. \quad . \quad . \quad . \quad . \quad (46)$$

Now consider the collision with reference to the K' system, moving with velocity v with respect to K, and let us choose v to be equal to V, so that, at the moment of collision, the bodies are momentarily at rest in the K' system. This can be so only if their original momenta in that system were equal and opposite; i.e. if

$$\frac{m_1'}{m_2'} = -\frac{V_2'}{V_1'}.$$

Unless, therefore, the mass of a body is inversely proportional to its velocity—and we know by experiment that it is not—this condition demands that $m_1' = m_2'$ and

$$V_2' = -V_1'. \quad . \quad . \quad . \quad . \quad (47)$$

Now substitute for V_1 and V_2 in (46) from (36), putting $V = v$ and using (47). We obtain

$$m_1 \cdot \frac{V_1' + v}{1 + \dfrac{vV_1'}{c^2}} + m_2 \cdot \frac{-V_1' + v}{1 - \dfrac{vV_1'}{c^2}} = (m_1 + m_2)v, \quad . \quad (48)$$

which reduces to

$$\left. \begin{array}{c} \dfrac{m_1}{m_2} = \dfrac{1 + \dfrac{vV_1'}{c^2}}{1 - \dfrac{vV_1'}{c^2}} \\[3em] = \dfrac{\sqrt{1 - \dfrac{V_2^2}{c^2}}}{\sqrt{1 - \dfrac{V_1^2}{c^2}}}, \text{ from (39).} \end{array} \right\} \quad . \quad . \quad (49)$$

Hence

$$m_1\sqrt{1 - \frac{V_1^2}{c^2}} = m_2\sqrt{1 - \frac{V_2^2}{c_2}} = m_0, \text{ say,}$$

or

$$\left.\begin{aligned} m_1 &= \frac{m_0}{\sqrt{1 - \frac{V_1^2}{c^2}}} \\ m_2 &= \frac{m_0}{\sqrt{1 - \frac{V_2^2}{c^2}}} \end{aligned}\right\} \quad . \quad . \quad . \quad . \quad (50)$$

It follows, then, that in order that the definition of mass contained in (41) and (42) shall be equally applicable to the K and K' co-ordinate systems, the mass of a body in each system must be a function of its velocity. m_1 and m_2, it will be remembered, are the masses of perfectly similar bodies, so that equations (50) may be considered as applying to the same body. The mass of the body at rest is clearly m_0. We have the same dilemma here that we noticed in connexion with length (p. 30), and we have to decide whether to call m_0 the mass of a body, and regard mass as invariable but not satisfying the laws of conservation of mass and momentum ; or to call the mass $\dfrac{m_0}{\sqrt{1 - \frac{V^2}{c^2}}}$, a

quantity which satisfies the conservation laws but changes with motion. We again choose the latter alternative, and to avoid ambiguity we call m_0 the ' rest-mass '. It would be a good idea to speak similarly of the ' rest-length ', but the term is rarely used.

One point should be specially noted, since the simplified treatment here given does not bring it to light. In equations (50), V_1 and V_2 represent the whole velocity of the body, in whatever direction it may be. Mass, even in its new form, is not a vector quantity, and the components

(M_x, M_y, M_z) of the momentum of the body, for instance, must be written

$$M_x = \frac{m_0}{\sqrt{1 - \dfrac{V^2}{c^2}}} \cdot V_x$$

$$M_y = \frac{m_0}{\sqrt{1 - \dfrac{V^2}{c^2}}} \cdot V_y \Bigg\} \quad . \quad . \quad . \quad . \quad (51)$$

$$M_z = \frac{m_0}{\sqrt{1 - \dfrac{V^2}{c^2}}} \cdot V_z,$$

where $\qquad V^2 = V_x{}^2 + V_y{}^2 + V_z{}^2.$

Experimental methods of measuring mass (other than by weighing, which, as we have seen, is an indirect process yielding a result which, apart from the general theory of relativity, can be identified with mass only on empirical grounds) should confirm equations (50), for they consist in principle of measuring the resistance which the body offers to a force. This is done by noting the *change* of motion produced in the body, and applying the conservation laws. Hence the mass measured is that which obeys those laws—namely, that given by (50). It is, of course, very difficult to obtain a wide range of observations at velocities such that $\dfrac{V^2}{c^2}$ becomes appreciable, but experiments on charged particles at speeds approaching that of light definitely indicate that (50), and not the pre-relativity formula (52), agrees with observation.

Conceptual Character of Mass. The purely conceptual character thus given to mass should be noted. The velocity of a body depends only on our caprice ; we can make it what we will by a suitable choice of a standard of rest. Hence the mass of a body also is a matter of caprice.   Our liberty ends, however, when more than one body come into consideration, for we must adopt the same standard of rest for both.   All physical relations between them then remain the same, however we exercise our choice, but those relations are not what we thought they were when we were restricted to phenomena at small relative velocities.

This purely conceptual character of mass is altogether incompatible with the old notion of the ' quantity of matter ' in a body.   This may be shown most strikingly by considering a co-ordinate system (the K' system in the above calculation) in which two equal masses approach one another with equal and opposite velocities, V', and come momentarily to rest on collision.   Before collision the mass of each is $\dfrac{m_0}{\sqrt{1 - \dfrac{V'^2}{c^2}}}$, a quantity which is greater than m_0, but at the moment of collision it is m_0.   Each body, therefore, has lost mass, yet it is a condition of the validity of our equations that the total mass has remained constant.   We shall see presently what has become of that which is ' lost ' : the point to notice at the moment is that, not only is mass not the ' quantity of matter ' in a body, but also it can exist in some form other than matter.

Mass in Pre-Relativity Physics.   It is interesting to repeat the calculation just made, using the classical transformation equation for velocity ($V' = V - v$) instead of that required by relativity.   We then have, instead of (48),

$$m_1(V_1' + v) + m_2(- V_1' + v) = (m_1 + m_2)v,$$

i.e.
$$m_1 = m_2. \quad . \quad . \quad . \quad . \quad . \quad (52)$$

The independence of mass and velocity, which we saw was not guaranteed by the equations of definition, (41) and (42), is thus required, according to classical physics, if we wish the definition to hold good in all systems of co-ordinates moving with uniform relative velocities. The demand is, of course, illogical, for we have no right to force mass to be invariant by employing transformation equations for velocity which are inconsistent with the invariance of laws of nature. So long, however, as it was believed that those transformation equations agreed with experiment, the independence of mass was a proper deduction.

The difference between the classical and relativity expressions for mass is clearly not important so long as v is small compared with c. As the velocity of a body approaches that of light, however, the mass increases, tending towards an infinite value when $V = c$. In other words, the resistance which a body offers to anything tending to increase its velocity tends to infinity as the velocity approaches c, so that this is a limiting value for the velocity of anything associated with mass. This gives us an additional sense in which we may regard c as a limiting velocity, but again, as on page 56, we must be careful not to give it this status without proper safeguards. We have no justification for saying that a velocity greater than c is meaningless. A material body cannot attain it, but a spot of light reflected from a revolving mirror, for example, may traverse a sufficiently distant surface with such a velocity, and, in wave mechanics, phase velocities of waves which exceed c are contemplated.

Transformation Formula for Mass. Equations (50) are, of course, not the transformation equations for mass. They apply only to a single co-ordinate system, and represent a condition which mass must satisfy in order that consistent transformation equations shall exist. The transformation equations can, however, now be obtained very simply.

Let m and m' be the masses of a body in the K and K' systems, respectively, and let the body have velocity V in the K system, and therefore velocity V', given by (35), in the K' system. We then have

$$m = \frac{m_0}{\sqrt{1 - \frac{V^2}{c^2}}},$$

$$m' = \frac{m_0}{\sqrt{1 - \frac{V'^2}{c^2}}};$$

whence

$$m' = m\sqrt{\frac{1 - \frac{V^2}{c^2}}{1 - \frac{V'^2}{c^2}}} = m.\frac{1 - \frac{vV_x}{c^2}}{\sqrt{1 - \frac{v^2}{c^2}}}; \quad . \quad (53)$$

by substitution for V' from (35).

It will be noticed that, notwithstanding the scalar character of mass, only the x-component of the associated velocity occurs in the transformation equation. This means that if two bodies, of equal rest-mass, are moving with the same speed in the K system, but in different directions, their masses, although equal in that system, will not be equal in the K' system.

If a body is at rest in the K system, then V_x vanishes, and the formula reduces to

$$m' = \frac{m}{\sqrt{1 - \frac{v^2}{c^2}}}. \quad . \quad . \quad . \quad (54)$$

Transformation Formula for Density. We can easily deduce the transformation formula for density, defined as mass per unit volume. We consider the simple case of a body at rest in the K system, and therefore moving with velocity $-v$ in the x-direction in the K' system. If W_0 and W' be the volumes in the two systems, then, since the dimension in the x-direction is to be multiplied by the factor $\sqrt{1 - v^2/c^2}$, we have

$$W' = W_0\sqrt{1 - \frac{v^2}{c^2}} \quad . \quad . \quad . \quad (55)$$

The relation between the masses is given by (54). Hence, for the densities, ω_0 and ω', we have

$$\omega_0 = \frac{m_0}{W_0}$$

$$\omega' = \frac{m'}{W'} = \frac{m_0}{W_0\left(1 - \dfrac{v^2}{c^2}\right)} = \frac{\omega_0}{1 - \dfrac{v^2}{c^2}} \quad . \quad . \quad (56)$$

In the more general case, in which the body moves with velocity $V_{x,y,z}$ in the K system, there is a 'contraction' in all dimensions. The corresponding formula may, however, easily be obtained from (35) and (53).

Energy and Mass. The mass m of a body moving with velocity V may be expressed by expanding (50), thus—

$$m = m_0 + \frac{1}{c^2}\cdot\frac{1}{2}m_0V^2 + \text{higher powers of } \frac{V}{c} \quad . \quad (57)$$

When the body is at rest the mass is m_0, and the effect of giving it a velocity V is therefore to add to its mass an amount $\frac{1}{c^2}\cdot\frac{1}{2}m_0V^2$, so long as higher powers of $\frac{V}{c}$ can be neglected. Now $\frac{1}{2}m_0V^2$ is the kinetic energy which is added to the body when it is given the velocity V. Hence the body acquires, at the same time, an increment of mass and a proportionate increment of kinetic energy.

Kinetic energy—and, in a lesser degree, mass also—is among those sophisticated conceptions mentioned on page 25, whose importance is not realized until physics has made considerable progress, and its significance depends entirely on the simplification which it introduces into the expression of physical relations. We now see that it is superfluous, because it is automatically introduced in the more complete expression for mass. We may therefore dispense with it as an independent conception.

If the attempts which have been made to reduce all energy to the kinetic form had been completely successful, we could have concluded at once that energy in general is simply mass expressed in terms of a unit c^2 times as small. Such a direct deduction cannot be made, but other considerations, for which we have no space, lead us to generalize our result and write, as a comprehensive equation,

$$E = mc^2, \qquad \qquad (58)$$

where E is the total energy and m the total mass of the body or mechanical system we are considering. We may then refer either to the mass or to the energy, according to convenience, the meaning being the same in either case. This implies that a body at rest has energy $m_0 c^2$: this has been called the ' energy of constitution ' of the body, and in theories according to which matter may be ' annihilated ', this is the amount of energy that is assumed to appear in some form or other when such an event occurs.

One example of the association of mass with energy in other than the kinetic form has already occurred in this chapter. We have seen that the sum of the masses of two elastic bodies when they come to rest at the moment of collision is less than the sum of their masses just before collision. We can now say that the ' lost ' mass is the mass usually expressed as ' potential energy of deformation '.

Another interesting implication of equation (58) is that a body which radiates energy must be regarded as losing mass. The energy radiated by the Sun, according to this view, indicates that the mass of the Sun is decreasing at the rate of about four million tons in each second.

Force. The measurement of force was prescribed by Newton in terms of the rate of change of motion (momentum) of a body on which the force acted. On the assumption that the mass of a body is constant, this has been interpreted as the product of the mass and the rate of change of velocity (i.e. the acceleration). With the more exact conception of mass which we now have, we see that it is necessary to return to the unmodified Newtonian measurement of force. We thus have

$$F_{x,y,z} \equiv \frac{d}{dt}(M_{x,y,z}) = \frac{d}{dt}(m.V_{x,y,z}) = m\frac{dV_{x,y,z}}{dt} + V_{x,y,z}\frac{dm}{dt} \quad . (59)$$

where m, of course, stands for $\dfrac{m_0}{\sqrt{1 - V^2/c^2}}$.

The first of these terms is the pre-relativity 'mass × acceleration', and the second, which was formerly neglected, must now be taken into account since dm/dt is not zero. Moreover, the two terms in (59) represent components which are not in general in the same direction; the former is in the direction of the acceleration, and the latter in that of the velocity. It follows that the force acting on a body is not, as we have thought, in the direction of the acceleration. It is in the direction of the 'change of motion', and there is a component of change of motion in the direction of the velocity because of the change of mass. In the case of a body (e.g. a cannon-ball) projected obliquely upwards, for example, the direction of the force is intermediate between that of the downward acceleration and the tangent to the path at the moment considered.

There are two special cases, however, in which the directions of force and acceleration coincide; namely, when the acceleration is respectively parallel and perpendicular to the velocity. In the former case—that of 'longitudinal' acceleration—the two terms of (59) can be compounded by simple addition. We may then consider the common direction of the velocity and acceleration, without resolution, and write

$$
\left.
\begin{aligned}
F = m\frac{dV}{dt} + V\frac{dm}{dt} &= \frac{m_0}{\sqrt{1 - \dfrac{V^2}{c^2}}}\frac{dV}{dt} + V\frac{m_0\dfrac{V}{c^2}}{\left(1 - \dfrac{V^2}{c^2}\right)^{\frac{3}{2}}}\frac{dV}{dt} \\
&= \frac{m_0}{\left(1 - \dfrac{V^2}{c^2}\right)^{\frac{3}{2}}}\frac{dV}{dt}
\end{aligned}
\right\} \quad (60)
$$

(If we then resolve the force along the axes, we must remember that V in the denominator retains its full value in each component, while V in $\dfrac{dV}{dt}$ is a vector quantity.) This expression would be applicable, for example, to the case of a body falling freely from rest towards the Earth's surface.

In the second case—that of 'transverse' acceleration—since the change of velocity is at right angles to the velocity, there is no numerical change of V, and therefore the mass remains constant. The second term in (59) vanishes, and we have

$$
F_{x,y,z} = \frac{m_0}{\sqrt{1 - \dfrac{V^2}{c^2}}}\frac{dV_{x,y,z}}{dt} \quad . \quad . \quad . \quad (61)
$$

In these two cases, therefore, we can still regard force as 'mass × acceleration', if we define mass in the former

as $\dfrac{m_0}{\left(1 - \dfrac{V^2}{c^2}\right)^{\frac{3}{2}}}$, and in the latter as $\dfrac{m_0}{\sqrt{1 - \dfrac{V^2}{c^2}}}$. These two

quantities were formerly known on this account as 'longitudinal mass' and 'transverse mass', respectively. As we have seen, however, it is only the second which preserves the law of conservation, and the other has long ceased to be regarded as significant. It is far better to retain the defining law of conservation than to preserve an inaccurate rendering of Newton's second law of motion.

Transformation Formulae for Force. The transformation formulae for force can be obtained from (59) by first writing that equation with all the quantities dashed, so as to refer it to the K' system, thus—

$$F_x' = \frac{d}{dt'}(m'V_x') = \frac{dt}{dt'}\frac{d}{dt}\left(m\frac{1 - \dfrac{vV_x}{c^2}}{\sqrt{1 - v^2/c^2}} \cdot \frac{V_x - v}{1 - \dfrac{vV_x}{c^2}}\right) \quad (62)$$

(with similar equations for the other components) and carrying out the differentiation, eliminating $\dfrac{dV_{x,y,z}}{dt}$ by means of the relation (59). The work, which is quite straightforward, is rather long, and we give the result, viz.—

$$\left.\begin{array}{l}
F_x' = F_x - \dfrac{V_y v}{c^2 - V_x v}F_y - \dfrac{V_z v}{c^2 - V_x v}F_z \\[3em]
F_y' = \dfrac{c^2\sqrt{1 - \dfrac{v^2}{c^2}}}{c^2 - V_x v}F_y \\[3em]
F_z' = \dfrac{c^2\sqrt{1 - \dfrac{v^2}{c^2}}}{c^2 - V_x v}F_z
\end{array}\right\} . \quad (63)$$

CHAPTER VII

ELECTROMAGNETIC MEASUREMENTS

General Field Equations. The general equations of the electromagnetic field, in terms of the so-called 'rational' units, are

$$
\left.
\begin{aligned}
\text{div } D &= \rho \\
\text{div } B &= 0 \\
\text{curl } E &= -\frac{1}{c}\frac{\partial B}{\partial t} \\
\text{curl } H &= \frac{1}{c}\left(\frac{\partial D}{\partial t} + I\right),
\end{aligned}
\right\} \quad . \quad . \quad . \quad (64)
$$

where D and E are the electric displacement and intensity, B and H are the magnetic induction and intensity, and ρ and I are the densities of electric charge and conduction current, respectively. There are also the additional relations,

$$
\left.
\begin{aligned}
D &= \varkappa E \\
B &= \mu H \\
I &= \sigma E
\end{aligned}
\right\} \quad . \quad . \quad . \quad . \quad (65)
$$

where $\varkappa$, μ and σ are respectively the dielectric constant, magnetic permeability, and electrical conductivity.

Transformation Formulae. The transformation formulae for these quantities, determined from the condition that the equations shall hold good in both the K and K' systems, have been calculated. The work is long, and we give only the result :

$$D_x' = D_x \qquad\qquad B_x' = B_x$$

$$D_y' = \frac{D_y - \frac{v}{c}H_z}{\sqrt{1 - \frac{v^2}{c^2}}} \qquad\qquad B_y' = \frac{B_y + \frac{v}{c}E_z}{\sqrt{1 - \frac{v^2}{c}}}$$

$$D_z' = \frac{D_z + \frac{v}{c}H_y}{\sqrt{1 - \frac{v^2}{c^2}}} \qquad\qquad B_z' = \frac{B_z - \frac{v}{c}E_y}{\sqrt{1 - \frac{v^2}{c^2}}}$$

$$E_x' = E_x \qquad\qquad H_x' = H_x$$

$$E_y' = \frac{E_y - \frac{v}{c}B_z}{\sqrt{1 - \frac{v^2}{c^2}}} \qquad\qquad H_y' = \frac{H_y + \frac{v}{c}D_z}{\sqrt{1 - \frac{v^2}{c^2}}} \qquad \Bigg\} \quad . \quad (66)$$

$$E_z' = \frac{E_z + \frac{v}{c}B_y}{\sqrt{1 - \frac{v^2}{c^2}}} \qquad\qquad H_z' = \frac{H_z - \frac{v}{c}D_y}{\sqrt{1 - \frac{v^2}{c^2}}}$$

$$I_x' = \frac{I_x - \rho v}{\sqrt{1 - \frac{v^2}{c^2}}}$$

$$I_y' = I_y$$
$$I_z' = I_z$$

$$\rho' = \frac{\rho - I_x\frac{v}{c^2}}{\sqrt{1 - \frac{v^2}{c^2}}}$$

One important point may be noticed immediately. Purely electrical quantities transform into a combination of electrical and magnetic quantities, showing that the division of the field into these two parts is not fundamental but depends on the system of co-ordinates used.

Once more it may be well to emphasize the conceptual

character of these transformation formulae. One might be inclined to think that the existence or non-existence of a magnetic force was something independent of conceptions, and that such a force could not be made to appear or disappear by a change of mind : if a magnetic needle were placed in the region of interest, its behaviour would surely indicate the presence or absence of a magnetic field, whatever co-ordinate system one chose to assume. It must be remembered, however, that the needle must be at rest when mapping the field, and if the co-ordinate system (i.e. the standard of rest) is changed, a needle formerly at rest must be regarded as moving. Allowance must be made for its motion, and the result will be that while no change whatever happens to the needle or the field by the change of co-ordinate system, the *interpretation* of the behaviour of the needle will be changed. If the needle takes up no definite orientation we say in one case that there is no magnetic field, and in the other that there is a magnetic field which the needle does not reveal because of the compensating field set up by its motion.

Invariance of Total Charge. It follows from the equations that the magnitude of a charge, e, is unaltered by a transformation of co-ordinates from the K to the K' system. For, if W be the volume occupied by the charge, we have

$$e = \rho W. \quad \ldots \quad \ldots \quad (67)$$

Now from (66) (since $I_x = \rho V_x$),

$$\rho = \frac{\rho'\left(1 + \dfrac{v V_x'}{c^2}\right)}{\sqrt{1 - \dfrac{v^2}{c^2}}}$$

$$= \rho' \sqrt{\frac{1 - \dfrac{V_x'^2}{c^2}}{1 - \dfrac{V_x^2}{c^2}}}. \quad \text{(See pp. 59–60.) . (68)}$$

Also, if W_0 be the ' rest ' volume, we have, from (55),

$$W = W_0\sqrt{1 - \frac{V_x^2}{c^2}} \; ; \quad W' = W_0\sqrt{1 - \frac{V_x'^2}{c^2}},$$

so that

$$W = W'\sqrt{\frac{1 - \dfrac{V_x^2}{c^2}}{1 - \dfrac{V_x'^2}{c^2}}} \qquad . \quad . \quad . \quad (69)$$

Hence

$$e = \rho'W' \equiv e' \qquad . \quad . \quad . \quad . \quad (70)$$

Force on Moving Charge. It is well known that a charge moving in an electromagnetic field experiences a force given in vector notation by

$$F = e\left(E + \frac{1}{c}[v \times H]\right) \qquad . \quad . \quad . \quad (71)$$

This is not deducible from the general equations of the field according to classical theory, and has therefore ranked as an additional postulate. The modifications introduced by relativity, however, remove the necessity for this, since, when the proper transformation equations are used, the force appears as a consequence of the change of co-ordinate system. This may be shown as follows.

Consider a charge e, at rest in the K' system and therefore moving with velocity v in the K system. We then have

$$\left.\begin{array}{l} v_x = v \; ; \quad v_y = v_z = 0 \; ; \\ v_x' = v_y' = v_z' = 0. \end{array}\right\} \qquad . \quad . \quad . \quad (72)$$

In the K' system the force acting on the charge is simply that due to the electric field, since the charge is there at rest. If F' be this force, we have therefore

$$\left.\begin{array}{l} F_x' = e'E_x' \\ F_y' = e'E_y' \\ F_z' = e'E_z' \end{array}\right\} \qquad . \quad . \quad . \quad . \quad (73)$$

The value, F, of this force in the K system is therefore given by inserting these values in (63), and using (66),. (70) and (72). In applying these equations we interchange dashed and undashed symbols by changing the sign of v where necessary, and note that $B = H$ in a non-magnetic medium. We thus obtain

$$\left.\begin{aligned} F_x &= eE_x \\ F_y &= e\left(E_y - \frac{v_x}{c}H_z\right) \\ F_z &= e\left(E_z + \frac{v_x}{c}H_y\right) \end{aligned}\right\} \quad . \quad . \quad . \quad . \quad (74)$$

This is a particular form of (71), and is easily seen to be the form which that equation takes by virtue of the fact that we have considered a particle moving parallel to the x-axis. Hence, generalizing this result for motion in any direction, we obtain the vector equation (71).

This, like the Fizeau effect, is an example of a physical fact which is already implied by our laws of nature when they are properly expressed, but which, when the old, unsatisfactory definition of length was used, seemed to need a special hypothesis to explain it.

CHAPTER VIII

TRANSITION TO GENERAL RELATIVITY

The Minkowski Four-Dimensional Continuum. The natural generalization of the relativity theory to cover all motion would seem to demand a still more complex substitute for l, in which acceleration was involved. We might then expect to transform away a gravitational field by transforming to axes moving with a particle set free in that field, just as we can transform away a magnetic field in the manner shown in the last chapter. No success has been obtained in this way, however, and the possibility that the transformation formulae of the special theory of relativity might hold without modification for relatively accelerated axes is ruled out by the fact that Newton's law of gravitation is not invariant to a Lorentz transformation. The required generalization has come by way of a re-expression of the special theory which is due to Minkowski.

In the Lorentz formulae (21), let us make the following substitutions for t and v:

$$\left.\begin{array}{l} t = \dfrac{i}{c}\tau \\[2mm] v = ic \tan \theta, \end{array}\right\} \quad . \quad . \quad . \quad . \quad . \quad (75)$$

where $i \equiv \sqrt{-1}$. τ may thus be regarded as a sort of imaginary time, and θ as an imaginary angle. The equations, after a slight reduction, then become

$$\left.\begin{array}{l} x' = x \cos \theta + \tau \sin \theta \\ \tau' = \tau \cos \theta - x \sin \theta \end{array}\right\} \quad . \quad . \quad . \quad . \quad (76)$$

with those for y' and z' unchanged. Now these are the equations of transformation from rectangular axes (x, τ) to another set of co-planar rectangular axes (x', τ') inclined at an angle θ to the first set. The Lorentz transformation is therefore equivalent to a rotation of axes through an angle θ, which is a function of the velocity.

If we include the y and z axes, along which there is no motion, we must regard them as remaining unchanged in the rotation, and we may then form a geometrical 'picture' of a four-dimensional region or 'continuum' ('space-time', it is usually called), in which the Lorentz transformation for a particular velocity v is represented by a rotational displacement of the x and τ axes through a particular angle θ. What meaning can be ascribed to this we shall consider directly, but we will first see how this way of expressing the special theory has led to its generalization.

It is easy to see from (22) that the Lorentz formulae lead to the result

$$dx^2 + dy^2 + dz^2 - c^2\, dt^2 = dx'^2 + dy'^2 + dz'^2 - c^2\, dt'^2. \quad (77)$$

This quantity is therefore an *invariant* for a Lorentz transformation. If we represent it by ds^2, and change t to τ by (75), we obtain

$$ds^2 = dx^2 + dy^2 + dz^2 + d\tau^2. \quad . \quad . \quad . \quad (78)$$

Pursuing our analogy of a four-dimensional space-time continuum, we see that if dx, dy, dz and $d\tau$ are the co-ordinate differences between two neighbouring points in such a continuum, ds may be called the distance between them. For, if there were only three dimensions, (78) would represent the distance between two points in ordinary space—that distance, as we know, being unchanged by any change of co-ordinate system. To avoid confusion, however, ds is generally called the *interval*, the word 'distance' being reserved for three-dimensional space. Geometrically, then, the interval is the analogue in four dimensions of the distance in three dimensions, and we can appreciate its invariance to a rotation of axes.

We thus learn to form a distinction between what, in natural laws, belongs to nature, so to speak, and what is contributed by us in expressing them. The orientation of the axes is a characteristic of our arbitrary choice of a standard of rest, but the interval, which is independent of that orientation, is objective and is unaltered by our expression of it. Any other similarly invariant function of the co-ordinates which we may succeed in finding may also be regarded as belonging to nature, and therefore representing something physically important.

General Relativity. It is at this point that the generalization is made. The continuum implied in (78) is regarded as representing the course of events in the physical world which it is our object to describe. We can describe it in various ways, corresponding to the various co-ordinate systems we choose to adopt. The Lorentz transformation corresponds to a change between co-ordinate systems whose relative velocity is uniform, but we may make more general transformations of co-ordinates corresponding to various relative motions and positions and scales of space and time and so on, without changing ds. Our measurements of space and time will be modified in a manner depending on the particular transformation we make, but the world measured, which is represented by ds and the other invariants of the continuum, will remain the same.

By the 'world' measured, we do not necessarily mean the whole physical universe, but any isolated mechanical system in which we may be interested—e.g. the solar system, the interior of the Earth, &c. How, then, does the description show that we are dealing with one mechanical system rather than another ? The answer is that this is shown not by a particular expression for ds^2, which corresponds to the choice of a particular co-ordinate system, but by the whole group of such expressions which can be obtained by making all permissible transformations of co-ordinates.

Let us suppose we start with (78). A Lorentz transformation will, of course, simply reproduce this equation with dashed instead of undashed co-ordinates, but if we make some other change of co-ordinates we may get a more complex expression. For example, if we put

$$\left.\begin{array}{l} x = x' \\ y = y' + \tau' \\ z = z' \\ \tau = \tau' - x' \end{array}\right\} \quad \cdots \cdots \quad (79)$$

we shall obtain

$$ds^2 = 2dx'^2 + dy'^2 + dz'^2 + 2d\tau'^2 - 2d\tau'\,dx' + 2dy'\,d\tau'. \quad (80)$$

According to the theory, this represents the same mechanical system described in terms of different co-ordinates, and any other expression for ds^2 which may similarly be obtained from (78) by a transformation of co-ordinates represents again the same system.

The general expression of the second degree, however, contains an infinite number of examples which cannot be obtained from (78) by such a process. If we take the whole collection of those which can, they represent a single mechanical system described in all possible ways, and the remainder represent other mechanical systems, each described in all possible ways.

We may therefore summarize the theory in the following way. The general expression for ds^2 is

$$\begin{aligned} ds^2 = {}&g_{11}dx^2 + g_{22}dy^2 + g_{33}dz^2 + g_{44}d\tau^2 + 2g_{12}dx\,dy \\ &+ 2g_{13}dx\,dz + 2g_{14}dx\,d\tau + 2g_{23}dy\,dz \\ &+ 2g_{24}dy\,d\tau + 2g_{34}dz\,d\tau, \quad \cdots \cdots \quad (81) \end{aligned}$$

where the coefficients are any functions (subject to certain mathematical restrictions) of x, y, z and τ. This represents all possible mechanical systems described in terms of all possible systems of co-ordinates. If the coefficients are all given specific values, then the equation represents a particular mechanical system described in terms of a particular system of co-ordinates. Any change in this expression for ds^2 which may be brought about by a mathematically permissible transformation of co-ordinates represents the same mechanical system differently described : with the precautions mentioned on page 35 we may say that it is the same mechanical system described by an observer in different physical circumstances. Any expression for ds^2 which cannot be obtained by such a transformation represents a different mechanical system.

We cannot here enter into the means by which the actual distribution of matter and motion in the system is obtained from the 'metric', or 'line-element', as the expression for ds^2 is called. We will only say that the rules for doing this show that the particular system represented by (78) or (80) is one which contains no matter or energy ; i.e. it is a region of empty space containing no fields of force. In such a region the only natural movements which can occur (exhibited, we may imagine, by test-particles introduced into it, the masses of such particles being negligible) according to Newton's first law, are those with constant velocity, and the only physically interesting transformations of co-ordinates are therefore the Lorentz transformations. It follows that the special theory of relativity is applicable only when the region in which the mechanical or electro-magnetic system exists is approximately free from matter or energy or gravitational fields—a condition which is realized often and closely enough to allow a wide application in physics but prevents the use of the theory on the cosmic scale.

Significance of the Four-Dimensional Continuum. In estimating the meaning of the generalization thus outlined, it is important to realize exactly what has been done.

In the first place, the processes of the special theory have been re-expressed in symbolic form, and the symbols used have themselves suggested the generalization. The Minkowski four-dimensional continuum is essentially metaphorical. When we represent a change in our standard of rest by a rotation of axes, we are leaving prose for poetry. We are doing what Shakespeare did when he represented the illumination of mountain peaks by the morning sunlight as jocund day standing tiptoe on the misty mountain tops. The representation is beautiful, and scientifically permissible if we do not forget that it is symbolic. We may extend it with profit if we bear this in mind, but at great peril if we do not. It is permissible, for instance, to look for the prints of the toes of day on the mountain, provided that what we expect to observe is, say, a patch of melted snow; but if we look for actual impressions of toes we are making nonsense of the whole conception.

Minkowski's metaphor has received both valid and invalid extension. The general theory of relativity is a valid generalization, as is shown by its agreement with and prediction of observations. But Minkowski himself began the invalid extension when he made his famous remark that henceforth space and time in themselves vanish to shadows, and only a union of the two exists in its own right; as though you could make the Sun disappear by representing one of its effects as a dancing boy. We have already traced the connexion between time and space—an association begun by our voluntary decision to measure time in terms of space. To symbolize that association by the image of co-ordinates in an imaginary continuum and then to endow the symbol with reality and call what is symbolized a shadow, is a proceeding which has only to be understood to be at once condemned.

Unfortunately the invalid extension has not stopped there. The metaphorical continuum is often called a 'space' in the mathematical sense (mathematical space is anything which may be described in terms of co-ordinates, and may have any number of dimensions. Thus we have 'configuration space' in statistics, and theories of 'abstract spaces' which have no connexion whatever with what we usually call space). A certain function of the co-ordinates used to describe a space is known as the 'curvature', because in the very special case in which the space does correspond to an actual physical surface, this function happens to measure the curvature of that surface. Now the curvature of the generalized four-dimensional continuum of Minkowski has a physical meaning, but it expresses that meaning symbolically. I have tried to translate the metaphor elsewhere * and have no room for it here, but it is important to notice the misunderstanding which has arisen. The 'space' which is the metaphorical continuum has first been wrongly identified with ordinary three-dimensional space which is familiar to everyone, and then the 'curvature' of the continuum has been identified with the curvature of a two-dimensional surface such as a sphere, and accordingly we are told that 'space is curved'. Thus a complicated set of misinterpretations of symbols, resulting in something which is meaningless and quite impossible even to imagine, has been presented as a great discovery of a physical truth. There is nothing at all unintelligible in relativity, though in the general theory there is, of course, difficult mathematics. All the inconceivable statements which have been made about it are the results of confusion of symbols with what is symbolized.

It is not always easy to avoid such mistakes if one wishes to make progress. As the general theory of relativity shows, even the most abstract symbol may be extended with great advantage, and it would be unwise to forego

* *Through Science to Philosophy* (Oxford University Press), Chapter XIV.

the advantage through fear of overstepping the bounds of legitimacy.   What is required is boldness in adventure, combined with care in estimating the significance of what is achieved.   A simple example will show how narrow is the line between true and false generalization.   A graph is a very well known symbolical way of expressing observations.   The ordinates and abscissae represent, by distances in space, things which are not necessarily distances. Take a nurse's temperature chart and a graph showing the variation with time of the height of a projected cannon-ball, for example.   The co-ordinates in the former case are temperature and time, and those in the latter, height and time.   Both graphs may be curves, but to say that a man's temperature has a curvature would be nonsense, while to say that the path of a cannon-ball is curved would not only be sensible but would actually be true if it was true of the graph.   It happens in the latter case that the abscissae, representing times, represent also horizontal distances covered, since the horizontal velocity (neglecting air resistance) is constant, and so, through an accident, the metaphor is capable of direct legitimate extension.   It is the business of physics to make use of such accidents, but it is no less its business to recognize that they are accidents and not natural necessities.

Philosophical Aspects of Relativity.   Relativity is essentially a physical theory, in itself no more and no less philosophical than any other such theory.   Nevertheless, it is probably of more importance to philosophers than any other department of physics.   This is not for the reason which in its early days made it appear metaphysical, namely, that it is concerned with space and time. We have seen that it is concerned with space and time in exactly the same degree as pre-relativity mechanics ; it has nothing to do with their ' nature ', but has simply modified the relations which have always existed between their measures.   The philosophical importance of relativity arises from the light which it sheds on the character of physical thought ; not from what it introduces into physics but from what it reveals as having always been latent there.

One aspect only of this revelation can be mentioned here ; namely, the better understanding which we now have of what is rational and what empirical in physics. In pre-relativity days it was possible for a physicist to be a naïf realist—as, in fact, most physicists, consciously or unconsciously, were. He could believe that he was dis-covering laws of a world of matter which was external to himself—laws which had nothing to do with his own thought and which simply described relations between objective qualities of matter which he could discover but could not create or destroy. To-day such an attitude is impossible. All such ' qualities ' of matter are seen to be concepts which we define for ourselves, for the one which we have chosen as fundamental (namely, length), in terms of which to express all others, is itself a function of an arbitrary quantity, v, which is at our disposal. More-over, such so-called objective ' qualities ' are not merely revealed as subjective concepts, but are also seen to be detached from matter. Even mass, the ultimate primary quality of a body, we have seen to be a solution of a pair of equations, and to be able to ' exist ' in other than material form. It is no longer possible for the physicist to maintain that his laws are laws of behaviour of matter.

This consideration has led at least one philosophically-minded physicist—Sir Arthur Eddington—to the belief that the laws of physics are purely rational ; that they could have been deduced, by a sufficiently powerful intel-lect, without recourse to experience, and that violation of them is impossible.* We cannot pursue this question

* See *The Philosophy of Physical Science*, by Sir Arthur Edding-ton (Camb. Univ. Press).

here, but it may be said that such a conclusion, whether true or not, goes beyond anything that can be deduced from the nature of physical laws as relativity has brought us to see them. It is true that Eddington's conclusion is based not only on the special theory, but also on the general theory of relativity and on the quantum theory; but, in the present writer's opinion, that wider ground affords no more solid basis for such a doctrine than does the special theory. What we are bound to grant (indeed, it can be read clearly enough now in the whole history of physics, but it needed relativity to make us aware of it) is that physical laws are not properly described as laws of behaviour of *matter*; it does not follow, however, that they are not essentially laws of regularity in *experience*. We must recognize that the ultimate data of experience with which physics deals are not the complex associations of sense-data and rational conceptions which we call ' material bodies ', but the bare sense-data themselves, and that the so-called ' qualities ' of material bodies are concepts which we form for the purpose of expressing the regularities which we find among these sense-data. In other words, we must draw the line differently between the empirical and the rational in our knowledge of the world. But that does not justify us in supposing that the rational part could be uniquely arrived at without experience, or that future experience may not show that it stands in need of amendment.

One point, already exemplified on pages 43–44, might be mentioned in this connexion—namely, that whether a law, stated in a particular way, is rational or empirical, may depend not on the facts of observation but on the concepts in terms of which we express those facts. Many of our present definitions would appear as empirical laws if we had defined the quantities taking part in them independently of one another. Nevertheless, however we choose our definitions, there is always a residuum, in any law which may be exemplified by experience, which cannot be created by pure reason.

Our most trustworthy safeguard in making general statements on this question is imagination. If we can imagine the breaking of a law of physics, then the law may possibly be falsified, and it is therefore in some degree an empirical law. With a purely rational law we could not conceive an alternative. We can imagine, for instance, that Jupiter may begin to move faster than the rate prescribed for it in the *Nautical Almanac*. We must therefore admit the possibility, however unlikely the occurrence might be, that it will do so and thus prove our present law of gravitation to be false. That law is consequently at bottom empirical. But we cannot imagine that the frequency of a wave of light will cease to be the quotient of the velocity and the wave-length, because we define the frequency in that way. The equation $\nu = c/\lambda$ is a rational law. Whatever observations we make, and however c and λ may change, the equation must still be true because ν is not an observable, but a quantity defined as c/λ. If an experimental determination gave another value, we should conclude that the experiment determined not ν but something else. This ultimate criterion serves as an anchor to keep us from drifting unduly in a perilous sea of thought.

INDEX

Aberration of light, 14, 17, 21, 50, 58
Absolute motion, 1
Acceleration, 7, 60
Airy, G., 14
Arago, 11, 12, 18

Clocks, 39
Composition of velocities, 55
Conservation of mass, 63, 65 68, 75
  of momentum, 62, 65
Co-ordinates, 31, 34
Curvature, 87

Definitions, v, 23, 38, 62, 63, 68
de Sitter, 17
Doppler effect, 3, 47, 56

Eclipses, 38
Eddington, A. S., 89, 90
Electromagnetic field equations, 76
Energy, 71
Ether, 6, 10, 14, 21, 52
Experience, vi, 2, 35, 46, 89, 90

Fitzgerald contraction, vi, 17, 20, 24, 30
Fizeau experiment, 10, 28, 57, 58, 80
Force, 73
Force on moving electric charge, 79

Frequency of light, 91
Fresnel, A., 12, 18

Galileo, 5
General relativity, 8, 81, 83
Gravitation, 7, 81, 91

Hypotheses, 52

Imagination, 91
Interval, 82
Invariance, 30, 57, 78, 81, 83

Kennedy-Thorndike experiment, 10, 18, 20, 24

Laws of motion, Newton's, 38 62
Length, v, 8, 23, 25, 29, 30
Lewis, G. N., 64
Light, 6, 47, 52
Limiting velocity, 56, 69
Longitudinal mass, 74
Lorentz, H. A., 17
Lorentz transformation formulae, vi, 42, 44, 47, 81

McCrea, W. H., v, vi
Mass, 62
Mathematics, vi
Matter, 90
Meaning, 1
Measurements, 25, 31, 67
Measuring instruments, 35, 40

Metaphysics, v, 31
Michelson-Morley experiment, 10, 14, 23, 42, 43
Miller, D. C., 20
Minkowski, 81, 86, 87
Momentum, 67
Motion, 1

Nature, v, 40
Nautical Almanac, 91
Newton, I., 5, 73

Observer, the, 34, 85
Oppolzer, 38

Philosophy, v, 88
Physical concepts, vi, 5, 25, 52, 67, 71, 78
relations, 25, 56, 78, 90

Reason, 89
Relative motion, 1
Relativity of position, 27
Resistance, electrical, 25
Rest-length, 66
Rest-mass, 66
Rotation of earth, 39

Sense-data, 90
Shakespeare, W., 86
Simultaneity, 45, 46

Space, 31, 43, 86, 87
Space-time continuum, 31, 82, 86
Standard of rest, 27, 30, 35
Symbolism, 86

Terminology, 30, 66
Thermodynamics, second law of, 4
Time, 19, 29, 33, 37, 86
Tolman, R. C., 64
Transformation Formulae :
    Acceleration, 61
    Density, 71
    Electric charge, 79
    Electromagnetic quantities, 77
    Force, 75
    Instants, 41, 42
    Length, 39
    Mass, 70
    Points, 33, 42
    Time-duration, 39, 45
    Velocity, 54
Transverse mass, 74

Universe, the, 2, 87

Velocity, 20, 28, 43, 52, 64, 69
    of light, 6, 13, 56, 57, 91

RELATIVITY

A VERY ELEMENTARY EXPOSITION

BY

SIR OLIVER LODGE, F.R.S.

SECOND EDITION

METHUEN & CO. LTD.
36 ESSEX STREET W.C.
LONDON

First Published . . June 11th, 1925
Second Edition . . 1925

PRINTED IN GREAT BRITAIN

RELATIVITY[1]

I FIND that at different times different subjects interest humanity, and the subject of special dominating interest changes from time to time. Half a century ago, or perhaps less, evolution was the word to conjure with. Now it appears to be relativity. And not only the mathematicians and physicists, but many of the philosophers, are dealing with that subject in a comprehensive manner; Lord Haldane in particular is trying to do for Einstein what Herbert Spencer did for Darwin —that is to say, to take a scientific idea, so far treated mathematically, out of the intricacies of physics, and spread it all over life, as the relativity of all knowledge.

In so doing the philosophers occasionally

[1] A Lecture to the Literary and Philosophical Society of Liverpool on October 31, 1921, reported by a stenographer from shorthand notes taken upon the evening of the address.

come to grief in their physics in a mild way, just as the physicists come to grief when they deal with philosophy. The subject is sort of betwixt and between, and it is quite easy to make it incomprehensible. Whether it is possible to make it comprehensible —well, that is what we have got to ascertain. As to relativity in general, the use of relative terms and the question of absoluteness about any of them, you know that nearly all our terms are relative. Take right and left. People tell you a shop is on the right-hand side of the street. There is no meaning in that. It depends on the way you are going along the street. But if you say right and left of the river, right bank and left bank, that has some meaning, but of course it is relative to the direction in which the river is flowing. Then there is fore and aft. That is all right with regard to a ship, but the ship may be turning round, and so it is not an absolute direction at all. It is a relative direction ; it may correspond to all points of the compass.

Take east and west. That has reference to the earth. Hence you might say that to all people on the earth it has the same meaning. In a sense it has, as when you say that Berlin is east of London and west of Petersburg. Otherwise it rather depends on where you are, when you talk about east and west, unless you are dealing with the abstract east and west. That, however, is relative to the earth, so it cannot be the same for all observers. Now, if a thing is not the same for all observers it is not absolute; it is shown thereby to be relative to something. When we can find anything that has absoluteness about it, it must be very important. Among these relative terms, it is of some interest to ask, Is there any absoluteness about any of them? Take up and down, for instance. Is there anything absolute about that? If the earth were flat, up and down would have a definite meaning for everybody, and the same meaning. But it is round, and up and down has different directions for different people. Up and down in New York is at an angle with our up and down; hence evidently it depends upon the place where you are. Up and down, if you are referring to a train on a railway, is relative to the capital city of the country. Up and down a mountain ; there is no mistake about that, but it is relative to the mountain.

There are a number of other terms I need not labour, such as far and near—it all depends on whether you have got a motor car or whether you have to walk ; high and low, strong and weak, heavy and light, dear and cheap—all these refer to something human. Then we come to large and small. We say a planet is large and an atom is small, but what do we mean by large and small ? What is our standard of size ? Have we a standard of size ? I think our standard is the human body. Anything much bigger than the human body we call big ; anything much smaller we call small, in general. There is nothing absolute about that, and I doubt if we can imagine a limit of bigness, a limit of size. The stars are enormous, far more enormous than most people know. Their size has been measured lately. The star Betelgeux, for instance, in Orion, that red star that is beginning to rise rather late at night now at this time of the year, has had its size measured ;

and if it were put in the place of the sun the earth could revolve inside it. Its bulk would extend to the orbit of Mars, far beyond the earth's orbit. Its size is enormous, but still limited, and there does seem to be a limit of size possible to a star. Then what about the universe? Is that infinitely big? We simply do not know.

But putting bigness out of mind, what about smallness? Is there a limit of smallness? Is the atom the smallest conceivable thing? The electron is very much smaller. When I was younger, the atom was considered the ultimate thing of which everything was built; now it is a bulky thing comparatively. This shows how little we mean by large and small. We are accustomed in physics to think of the atom as quite large; the electron is the small thing, as small as a flea is to this hall when compared to the atom. Well, is that the end? Is there anything smaller than the electron not yet discovered? Is the electron small, absolutely small, I mean in the eye of Deity, not in the human eye, of course? We never see such things. We cannot see the atom; it is far too small for us to see, even with the most powerful microscope, because the waves of light are too big. We associate size with a certain complexity or possibility of complexity. We say a planet may have any number of things on it. Is such a thing as that possible to the electron, or is it too small? I do not know; I do not think anybody knows.

Then is there no absoluteness about any of these terms? Yes, strangely enough, about hot and cold. When we say a thing is hot we generally mean that we do not want to touch it. It is hotter than the human body, and when we say it is cold it is colder than the human body. Is there an absolute coldness? There is no absolute hotness. The sun is the hottest thing we know, except some of the stars, which are now believed to be thousands of times hotter. But there is an absolute coldness. We owe the determination of the absolute zero of temperature to Lord Kelvin chiefly. There is an absolute zero, the same for everybody, not only on earth but throughout the universe—one absolute zero of temperature, which is about 500 Fahrenheit degrees below the ordinary Fahrenheit zero. It is known with some accuracy; it is known within a degree, and experimenters have got to within two or three degrees of it, by means of liquid helium. We have not quite attained the absolute zero, but we know where it is, and it was calculated long before it was got anywhere near experimentally.

Now, how can there be an absolute zero of anything? Well, just consider what heat is. It is the irregular jostling of the molecules of matter. When the molecules of matter are vibrating or moving rapidly among themselves —not all together—the body is said to be hot. Heat is that motion, nothing else. If you slow them down so that they are more sluggish, the body is cooling, getting cold. Slow them down until they stop—that is absolute zero. You cannot slow them down any more; you have got to the zero when you have taken all the heat out of the body. Of course, if you take a man's capital away he can go lower— he can get into debt; but that is not possible with heat. It gets down to zero and then stops. But you might say, " Is not the thing moving ? " Yes, it is moving as a whole— locomotion. It may be moving, but that does not matter ; the common motion is not heat. Heat is the irregular jostling of the ultimate particles, and when that stops the body is absolutely cold ; it is at the absolute zero of temperature. And at that temperature it has remarkable properties. It is a perfect con- ductor of electricity ; so that if you start an electric current it goes on. There is nothing to maintain the moon's motion round the earth, but there is nothing to stop it. The same with an electric current in a body at absolute zero ; it goes on because there is nothing to stop it.

Now, I come to the question of Time. Take the words " sooner or later " or " before and after " or " past and future." Is there any absolute meaning for those, or are they relative terms too ? At first sight one would say that the past was past, that the future was future, and that the present was the mere slice bounding the two—an infinitely thin partition as it were between the past and the future, advancing, leaving more of the past behind, encroaching on the future; and that we live in that slice of " present." Well, there may be animals which do live in the slice of " present," and have no memory and no anticipation. We are not in that predicament; we do look before and after. But there are certain conditions which have led relativists to hold the dogma of the relativity of time, to say there is nothing absolute about time, that the time for different observers may be different; not merely a difference like difference of longitude, but in a way dependent upon the motion of the observer. Now, here I must explain that I am not a full-blown relativist. A full-blown relativist might not agree with all I say; but I want to represent the case fairly, though the relativity of time is not an easy subject, even to those who fully believe in it.

There is a difficulty about simultaneity. When two things happen, can we tell if they happen at the same time or not? At first sight you may say, " Why, yes, I can see them happen. I know if I do something here, something else happens at the same time;

I see it." But that does not allow for the time the light has taken to come. Well then I will employ the telegraph, and if a thing happens in New York and I have it telegraphed here, say by wireless or any other method, I can tell when it happens, and can be sure that it happens at the same time as something else. But you have to allow for the telegraphic delay, which of course is very small. The time occupied is the same, or practically the same (if you have a perfect method) as that required by the velocity of light. And there is a real difficulty about determining simultaneity when you come to experiments of great accuracy. Suppose you want to determine whether the observed velocity of light depends on direction, the direction of the motion of the earth. You may send a beam of light and telegraph its arrival. The beam of light takes a certain time to go, and the telegraph takes a certain time going back ; so the two going in opposite directions neutralize each other. Whether you use a beam of light to tell you of the arrival, or whether you use an electric wave, comes to the same thing ; they travel at the same rate. The reason is that they are both transmitted by the ether.

Moreover, the present moment is more than a slice. There is all that is happening at different places at about the same time, places at a distance. For instance, take the case of things happening at a great distance. In 1901 a new star burst out in the heavens, in the constellation of Perseus, I think it was. When did that happen? When you saw it? You know well that it did not happen when you saw it. It happened a good time ago, and certain circumstances connected with that star enabled people to calculate, with surprising accuracy, that it happened in the year 1603, just about three centuries before. When you saw that star you would say, " There is a new star now." Well, is it new " now "? It depends what you mean by " now." You see it now. The messenger which brought the news of the new star was light, and we know of no quicker messenger. Had it been any other messenger, such as sound, we could not have heard of it yet.

PAST, PRESENT, AND FUTURE

It is obvious and simple enough that the past controls the present; but intelligent beings are controlled also by the future. I sometimes think that that is the difference between life, especially animal life, certainly human life, and the inorganic world. The inorganic world—the atom, the matter—is pushed from behind. It is controlled entirely by the present and not by the future. But that is not the case even with an animal. A dog is controlled in his actions by his anticipation of dinner. He, too, looks before and after. He has some memory and some anticipation, just as we have more memory and more anticipation. Take for instance an eclipse. It is going to happen; it has not happened yet; you will not see it until things have had time to travel. That eclipse has caused an expedition to start now; at a certain date it started in order to see the eclipse. That is a case of anticipation. It caused the fitting out of a ship, the collecting of a number of instruments, and the travelling to a place where the eclipse would be visible after the lapse of a certain time. In that sense the future controls the actions of the present. It is only one example of the fact that our actions are largely controlled by the future. The inorganic world is wholly controlled by the past and present. To be influenced by the future is a sign of advance.

Now, a great deal can be said about the relativity of time, but I must be satisfied with saying this : that in relativity you have to consider different ways of dividing up space and time. It might be a common mode of expression to say that the French Revolution occurred so many hundred of miles and so many years away. Distance and time can be put together. Time and distance are related. For instance, you can say truly that a kilometre is ten minutes' walk. Or, again, if you are at York, you can say that London is 200 miles away, or if you are going by train you can say it is four hours away. You very often use time as a measure of distance.

It depends on the vehicle you are thinking of. It is velocity that unifies space and time, and

2

you can practically use time as another dimension of space ; not exactly as a dimension of space, but treated much as if it were. It is sometimes called an imaginary dimension of space. You see there are three dimensions of space. There is right and left, fore and aft, up and down.

Those are the three dimensions of space, and what room is there for a fourth dimension ? I wish I could draw a fourth dimension on the blackboard, but I cannot. How did I get over it when I drew three ? Only by a perspective dodge in which you acquiesced. I did not really draw even three. If you are to have time in the diagram as well, you must dispense with drawing one of the space dimensions. But whether you can draw it or not, you must imagine it ; you must imagine progression in time. I cannot draw it, and I cannot tell you what it is doing, because it is an imaginary figure and it may be changing in time. It may be changing as an expanding circle ; it may be changing even as a shrub ; or as a seed which, beginning as an acorn, grows into a big tree. What is it diagrammatically doing ? It

is advancing in the time dimension. Here is
a thing which you will admit is the same
thing, it has got an identity, just as the tree
has, but it can go through all sorts of changes
as it advances along the line of time, and then
it can decay. Somewhat in that way the
motion of planets, motion of anything, can be
treated. You can speak of it as motion in a
plane, or you can speak of it, if you like, as an
advance in another dimension of space ; and
if you have already got three dimensions, as
we have, in length, breadth and thickness,
then time must be a fourth dimension—not
accessible as a dimension, but imaginable, as
if we were going through a process of development. It is development, evolution ; development by travelling along the inexorable stream
of time. To us it is inexorable. We cannot
hurry it or slow it or stop it. Whether there
is anything absolute about that flow of time—
well, that is the question. Is it a human
limitation, or is it a Divine reality ?

You see we are getting into philosophy and
metaphysics now. I am not trying to give you
the answers to these questions, but to indicate

the kind of things meant when we talk about time being a fourth dimension, and the way in which time can be thus represented and thought of. One of the things that relativity asserts to be absolute is the completely specified interval between two events. People may differ as to how far apart they are. One will say " so much space and so much time," and another will divide up the space and time differently. Different observers, according to the theory of relativity, will split up the interval in different ways. They will say so many miles and so many seconds separate the Coronation of George V from, say, the death of Charlotte Corday. They may not agree about the miles, and they may not agree about the seconds, but they will all agree about the interval compounded of the two ; which is absolute, an invariant, that is, a thing that remains constant and independent of the observer ; the space-time interval is absolute. When relativity admits the interval between two events as absolute—the same for all observers, no matter how fast they are travelling, nor where they are—it has important consequences. That is one basis of the mathematical theory.

Now, why do relativists claim that all the separate spaces and times depend upon the observer and are not absolute? To explain that, I must take another pair of terms, quick and slow. When we say a thing moves quickly—say a cannon ball moves quickly and a snail moves slowly—is there anything absolute about that? Is a cannon ball really quick? Is a snail really slow? It depends upon our estimate of space and time. Ordinary motion, as we know it, is certainly relative. You walk about on the deck of a moving ship and there is a bewildering turmoil of relativities. There is the motion of yourself relative to the ship, there is the motion of the ship relative to the earth, there is the motion of the earth relative to the sun, and there is the motion of the sun relative to the stars. At what rate are you moving? At what rate are we moving now? I know we are moving nineteen miles a second, because that is the rate at which we are going round the sun. In the time taken for a pin to drop from the ceiling to the floor we have travelled nineteen miles. We do not look like it; it is not obvious; we think we are at rest, but we are not. We may be going very much faster, since the sun is moving too.

Is anything at rest? Motion is relative as far as we have ascertained. We do not know what rate we are moving nor where we are going. We do not know the direction nor the magnitude of our direction at this moment. We have no idea. You may say you have some idea, that you are moving with the earth, that the earth and you are moving round the sun and the sun is moving with reference to the stars. Yes, with reference to the stars; but what are the stars doing? What do you mean by the stars? You mean our visible cosmos, what we can see with a telescope, but that is not the whole. It is now thought that our system of stars, the Milky Way, our cosmos, is but one of many. Far away in the depths of space there are others, called Island Universes—other cosmoi, other Milky Ways, other systems of stars—and some of those are moving at a terrific pace, 200 miles a second.

What is our pace ? We do not know. But what do you mean by moving ? We can move with reference to the walls of the room or with reference to the earth, but what do you mean by moving, absolutely? What is your standard of rest ?

Here is where I differ from relativists. They would say, " You have not got a standard, and to talk of absolute motion is meaningless." I would say, to talk of absolute motion with reference to nothing at all, is meaningless, but I think that we have a standard, and that that standard is the ether of space—the medium in which we are moving, the medium which extends throughout the universe, a medium which it would be absurd to suppose was in locomotion. I say that is our standard of rest for all practical purposes ; and they would agree, if it exists, but as to that they differ among themselves and do not say much about it. They are quite reasonable about it, but some of their early writings make people think they have abolished the ether. Eddington does not say that, and Einstein does not. Eddington told me he had asked Einstein in

Berlin recently, who said, " No, I have no objection to the ether ; my system is independent of the ether." That is all right ; I agree ; but that ignoration does not abolish it. When Laplace was asked by Napoleon in reference to his System of the World " Where is the Deity ? " he replied, " I have no need of that hypothesis." His system had no need. If he was always to be invoking the finger of God to regulate the planets it would indeed be a poor system ! He had to explain their motions on simple mechanical principles ; and that is what he did.

But that does not exclude the Deity from the universe. It simply means He is ignored ; He is not essential for the mathematical theory. So it is with the ether. They ignore it because it is not necessary to their system, and they are quite right. If we differ, it is because they prefer to say absolute motion has no meaning, whereas I should say that absolute motion has a meaning with reference to the ether, but that we have not yet ascertained what that meaning is. In other words, we have not yet ascertained what that motion is.

Shall we ever be able to? Here comes the point—and this was the beginning of relativity —the proof, or shall I say the failure, the failure to find any motion through the ether. If you try to ascertain how quickly you are moving through ether, what do you find? Many people have tried to find it in the last half century, and they have completely failed. A great deal is known about the attempts made to measure or even detect the virtual stream of ether in which we exist. If the earth is moving through the ether it is the same as if the ether is streaming past the earth, relatively. You can illustrate that by reference to the air. It does not matter whether you are rushing through the air or the air rushing past you.

People have asked, " How does the earth manage to move through the ether? Does the ether stream through the earth like wind through a grove of trees? " They thought it must, but when they tried to discover the process they failed. They could not find any phenomena that depended on that. Then there was the famous experiment of Michelson, so often mentioned that I suppose you are tired of it, but he is a great experimenter, now or recently at the University of Chicago. He was partnered in this classical experiment by Morley, and they thought they could find our motion through the ether by its effect on the velocity of light. They said, " If we are living in a stream of ether, light must go slower against the stream than with it." They devised an ingenious method for measuring the velocity of light in different directions. They had to send it to and fro. The simple thing would have been to measure the velocity first one way and then the other way, but that you cannot do because you do not know when it starts. You can send it across the stream, and simultaneously along the stream ; but you must send it to and fro in either case.

Now, you can time those actions with enormous accuracy, and although there is compensation in the coming back, the compensation is not complete. I do not know whether it is obvious to you which would take the longest, to go a mile with and against a stream, or to go a mile and back across the stream, but if you do the arithmetic of it you find it takes longer to go with and against. You are assisted with and retarded against, but the coming back takes longer and allows more time for retardation. You are not helped or hindered in going across the stream ; at least, you are not hindered very much. That experiment of Michelson's apparently ought to have shown that we were living in a stream of ether, and it did not. It showed nothing. That was the beginning of the trouble. That is the experimental foundation for all this relativity. The velocity of light appeared to be the same whether going with the ether or against it, consequently it seemed to say, " There is no motion through ether at all." So some people went further and said, " You may just as well say there is no ether at all." But to say that, they could not have thought what light was, because they could not have waves of light if they did not have a vehicle or medium for them. All it really proved was that the virtual stream of ether, depending on the earth's motion through it, did not show itself.

Why did it not show itself ? Was it because

the earth carries the ether with it, or carries some ether with it, so that near the earth it is stagnant ? That would explain it, but then there was the experiment performed by me at the University College here in Liverpool a quarter of a century ago, when I was Professor there, which puts that out of court. I whirled steel disks at a great rate till they nearly burst, and sent light round and round between them, that way round and this way round, and compared the time taken to go round *with* the disks with the time required to go round *against* the disks ; for if the ether had been carried round with the disks the beam one way would have been accelerated, and the other way retarded. There would have been a small effect. There was none. The ether was not carried round with disks. This proved that ether and matter are mechanically independent of each other, there is no friction ; matter moves without the slightest resistance, and its motion does not affect the velocity of light in its neighbourhood. The possibility that the Michelson and Morley experiment could be explained by the carrying of some ether along with the earth was clearly disproved. Was there another explanation? Here were two experiments, both undeniable; nobody controverted either of them. They seemed to be contradictory.

The explanation was suggested by my friend, Fitzgerald of Dublin. We were discussing this, and he said, "Well, I believe the thing which holds his mirrors (with which the experiment was made) shrinks when it is moving." The starting point and rebounding point of the light were on a great slab of stone. If there were a stream of ether it would be flowing through this stone. Light ought to have taken longer to go to and fro along the stream. It did not. Why? Because that stone shrank longways, and the contraction made it a shorter distance, so that that longitudinal beam is at an unfair advantage as regards the transverse beam. If the shrinkage occurred, the beam of light might take just the same time as the one that went across. We considered it, and perceived that it would do what was wanted; and soon afterwards H. A. Lorentz, of Leyden, the great Dutch

Professor of Physics, carried it a little further. He took it up, or started it independently, and showed that on the electrical theory of matter, shrinkage ought to occur, if matter was composed of electrons. It ought to occur, and, calculating the amount of shrinkage, he found that it just compensated and neutralized the Michelson-Morley experiment.

You may imagine the carefulness of the Michelson-Morley experiment when I say that the amount of shrinkage needed to counter-balance, to compensate, the retardation, is only about three inches in the whole diameter of the earth. The earth is eight thousand miles in diameter, but if it is moving along, in that one direction it is three inches shorter, and when it comes round to the other direction it gets its three inches back again. Hence a relativist will tell you that if I hold a stick so, it is one length; while if I hold it so, it is a trifle shorter. The shape and size of bodies change according to their position. A relativist would tell you so; and I am ready and willing to tell you so too.

How can I tell? If I measure it I shall not

find out, because the measure shrinks too. How do I know that though I am 6ft. 3in. if I stand up, yet if I lie down I shall change—get a bit longer or shorter, whichever it is? Do I know how much? I don't. It depends how fast we are moving through the ether. I do not know how fast we are moving. Very well, then, you do not know how long you are. We live in a queer world of ignorance, and there is no mode of testing it; so that we not only have relativity of motion, but of size and shape. A sphere is not a sphere. The effect is small, or believed to be small, because the Michelson-Morley experiment only tested the motion round the sun, not through space. We could not test that. The motion round the sun changes in six months, but the motion of the sun through space you cannot change. You can make no experiments on that; it is hopeless. You cannot measure that; therefore we do not know what size we are, or what size anything is. Everything is relative; not only time, but velocity, motion, size, shape, mass, even matter; that is a consequence of the electrical theory of matter.

A pound of matter would be rather more, if it were moving quickly, than if it were not moving quickly ; so that mass is not constant, as Newton thought it was. Relativity says the same. It is consistent with relativity, but it is a sequence of the electrical theory of matter. My view is that the ether affects all these things. Motion through the ether is changing the length and size of the body, the shape of the body and its mass. Ordinarily there is no means of ascertaining these things, and hence they are all relative—nothing absolute about them. They may differ for different observers, and whether they have any absolute meaning, a relativist would say he did not know.

But is there nothing absolute about velocity ? Why may man not travel through the ether at the rate of a million miles a second ? Here we come across something new, something absolute ! Curious ! A relativist would admit it. There is a velocity—he would not call it a velocity in the ether ; he might say a velocity in space ; there is a velocity which you cannot exceed. If you try to move at 180,000 miles a second the ether will just let you go ; it will get out of the way, but with such reluctance that you find it extremely difficult. If you try to go 190,000 miles a second, it will not get out of the way, and you cannot go. A bullet cannot go through air quicker than the velocity of sound proper to the heated air in front of the bullet. Sound has the velocity at which air will get out of the way. When dynamite explodes, the air declines to get out of the way. The air is made to get out of the way, but then the building is made to get out of the way too. One resists as much as the other. The velocity of light is the velocity at which the ether will not get out of the way, and consequently no piece of matter can move quicker than that. That is a maximum velocity. Hence to say that the stars may after all be revolving round the earth once a day, and that we cannot be sure that the earth is rotating on its axis, is essentially absurd ; not only absurd but demonstrably false.

There appears to be something absolute in the velocity of light. And this is a most remarkable conclusion of the relativist. He would say—I would not say it, or rather I would like it to be confirmed by experiment—that whether there is a stream of ether or not, light takes the same time going with the stream as against it. Suppose you are travelling to meet the source of light, surely you will get it quicker than if you sit still, or if you run away. They say no : whether you are going faster or slower, you will get it at what seems precisely the same time.

So that there are two things absolute, the interval between two events, when you take both space and time into account, and the speed at which light moves. How does that come out of relativity ? It comes out of the composition of motions. I cannot stay to explain fully how we get the composition of motions, but if you have two velocities they compound together. Supposing you are in a boat on a river ; you are going, say, four miles an hour and the river is flowing three miles an hour in the same direction. What is your velocity with reference to the earth ? What is your actual velocity ? Your velocity is four miles relative to the river, that of the river is three miles relative to its banks, so you get seven miles altogether. If you are going in the opposite direction you get one mile. Is not that common sense? If you have two velocities u and v you compound them into $u + v$, but when you apply the relativist doctrine, mathematically, taking into account all that I have tried to sort of skim the cream off, that is not the law of composition. The resultant velocity is not that, but

$$u + v \text{ divided by } 1 + \frac{u\,v}{c^2}$$

where c is the velocity of light. This un-expected denominator is $1 +$ the product by the square of the velocity of light. That fraction of $u + v$ gives w, the resultant velocity. That denominator is introduced by the theory of relativity, it is introduced by the Fitzgerald contraction, by all the different things I have been explaining, and I admit it is there. It is a very curious thing, it is a very odd formula for the velocity. If the velocity c were infinite, the whole thing would be common sense again —the resultant velocity would be $u + v$

simply—but as the velocity of light is not infinite, there is a very small correction which has to be applied ; which, strangely enough, has to be applied in actual practice when things are moving quickly enough. Some of the planets are moving quickly enough. Mercury is moving quickly enough, and it affects the motion of Mercury slightly.

Now go further and suppose that I am compounding something with the velocity of light itself ; instead of only the motion of the earth relative to the sun, and the sun relative to the stars, or instead of any other two motions that you can think of which you might take as u and v. Take my experiment with the revolving disks. I was trying to modify the velocity of light by compounding that velocity with another one, that of the disks, or the velocity of the ether between the disks. I was looking for a velocity $c + v$, trying to compound c with another velocity v, somewhat as in the Michelson and Morley experiment. I was sending light down the stream and up the stream, aiming at $c + v$ and $c - v$; trying, in fact, to see if the velocity of light increased up stream and diminished down stream. Neither they nor I found anything. Why not ? Because that is not what could have been found. Look at this equation. It expresses the new law for compounding velocities, and algebraically the result is c.

$$w = \frac{c + v}{1 + \dfrac{cv}{c^2}} = c.$$

They did not know it, I did not know it, but that is the law of composition according to this formula, when one of the velocities is c and not u. Work that out algebraically. Give it to your boy and he will tell you the result is algebraically c. It comes out the velocity of light and nothing else. You have tried to increase the velocity of light by adding v to it, but you cannot. It is unchangeable. Hence, the experiments were all bound to give negative results, without any talk about the ether, without any talk even about the Fitzgerald shrinkage, because of that law of composition which is the law appropriate to relativity, and its curious treatment of time.

I am coming to the end of my programme, though I have still got to introduce gravitation. When we introduce gravitation all manner of other things happen. You begin to doubt Euclid, and to talk about the nature of space; relativity is supposed to do away with gravitation. When you come to look into the matter, as to what you really observe, instead of only what you think you observe, you find a difficulty. You think you observe an attraction of one body for another. The earth attracts the moon. How can it attract the moon when it is not there? There is a great distance between the bodies. How can any body act directly at a distance? Newton knew it could not, but he did not know enough to explain how it happened. He could surmise, as we can and do, that both the earth and moon act on ether, and that the ether presses them together. But statements like that are of no value until they are worked out. We can safely say that the moon moves exactly as if the earth attracted it but the apparent attraction has still to be explained.

Einstein's is an attempt to work it out,

using different language. He would say: Here is a particle moving by itself in empty space. Here I put in its path not exactly an obstruction but a curvature, a pucker. Let the particle be moving in a sheet like a stretched plane, and let us make somewhere in that plane a pucker, and let the particle have to go near that pucker, which we will call a mountain. Suppose you want to go the easiest way, you won't go like the land crabs or some animals do, straight up the mountain or the house or whatever the obstruction is —up and over and down. You will prefer your path diverted, you will try and go round it. Your path will become curved, to get as little of the pucker as possible. You will stretch the apparent length of your path to get an easier, shorter-time path. Your path will be curved with reference to this pucker. But you might also express the motion by saying that the moving thing is apparently attracted by the pucker and so curved round. It is something like the hyperbolic orbit of a comet attracted by the sun. Space seems warped.

What is that warp ? A warp in space has caused the path of a body to be curved in order to get the line of least resistance. No longer would you call it an attracting force. You would simply say that the path is the effect of a warp in space. What is the warp due to ? You might say, to an atom or mass of matter at its centre. You might claim that matter warps the space all around it, and accordingly that the gravitational behaviour of bodies is as it is. We live in a warped space.

Now, is it right to say that matter causes the warp, or that the warp is matter ? A full-blown relativist would say that that is what you mean by matter. These warps rise in the centre not only to a pucker that can be got round, but to one that *must* be got round ; so that if you try and go through the centre you are up against the impenetrable. You cannot get through the centre.

The impenetrability of matter follows, therefore, as well as the attraction of matter. I do not suppose I have made that clear at all, but still I have indicated roughly the kind of way in which these warps in space simulate and replace the effect of gravitation. Only to me a warp in empty space is meaningless. An effect on the ether is full of meaning, and I believe the sun and all the planets do really affect the ether in such a way as to produce their actual paths, which we may likewise attribute to a pressure on them all towards the sun. You may as well call it a warp as anything else; and by calling it a warp you can, for the time, avoid the necessity not only of gravitation but of matter itself. Everything then becomes reduced to geometry, and Euclid's propositions are not strictly true. In the warped space or ether you have a different kind of geometry.

Now, strangely enough, the geometricians of the past had invented a hypergeometry that was not Euclid's, and that seemed to have nothing to do with anything but imaginary and ideal laws. Einstein had the genius to perceive that this hypergeometry would do what he wanted in the physical real world. By using that geometry he could work out the whole of the universe, so to speak, on geometrical lines; dispensing with physics, force, matter—with any of those things that we have lived on—and reducing it all to pure mathematics. It was a *tour de force*. It is a wonderful achievement, very brilliant, and I do not wonder mathematicians are enamoured of it. But the end is not yet, and we shall come out into common sense later on, with the addition of those great and notable discoveries which have followed from this method of treatment. For, mathematically considered, relativity is a splendid instrument of investigation, a curiously blindfold but powerful *method* of attaining results without really understanding them. There have been several of such methods—second law of Thermodynamics and others—but they ultimately have to be explained by physics. They are not a substitute for physics, they are not a philosophy. If pressed unduly you can manage to express things rather absurdly, but the method is a way of arriving at real results and of dealing with abstruse and hidden problems. Relativity is not a replacement but a supplement of Newton.

Lastly, consider for a moment the relativity

of human knowledge. Eddington says, towards the end of his book, *Space, Time, and Gravitation* :

The theory of Relativity has passed in review the whole subject of physics. It has unified the great laws which by their position hold a proud place in knowledge, and yet this by itself is only an empty shell. The reality is in our own consciousness. There are mental aspects deep within the world of physics. We have only regained from nature what man has put into nature. Everything is relative to human perception.

This may be understood, or misunderstood, as meaning that there is no objective reality at all, that things are as it were brought into existence by our conceptions of them, that a subjective existence is all the existence they have. Any such interpretation as that I repudiate. Our perceptions enable us to disinter from nature some part of what is already there—and which we certainly did not put there—but the phenomenal aspect which reality assumes to us, in other words its appearance, does depend on our modes or channels of perception and on our interpretative human mind. Objective reality exists, but it is we who interpret it. The universe is incapable of being completely comprehended by any finite being, it must be interpreted; and the way we interpret it depends on ourselves and on our faculties. Absolute reality might presumably be apprehended and formulated and perceived in a great number of different ways; we apprehend it in a human way, and our science must be conditioned by the human mind; it is therefore bound to be relative to the human mind. But the human mind is not a constructor of nature—only an interpreter. Objective reality exists, and makes an impression on us. The impression it produces depends on what we bring to its perception. Surely it is the same not only in Science but in Art. An artist perceives, or in a sense discovers, and displays his vision of reality to those who have senses to appreciate—not senses only but an adequate mind too. For example, a man perceives one aspect of a work of art, an animal perceives another—in so far as it perceives it at all. In that sense, and in that sense only, we get from nature what we put into it. We do not doubt that man sees it more truly than the animal. How God perceives it, or what it is in ultimate reality, we do not know. Our interpretation is relative to our own consciousness, even to our own individual consciousness ; but there are levels of consciousness, and science seeks to raise our conceptions above what is merely individual, and aims at universal truth—truth acceptable to all humanity.

And so, concerning all the discoveries which have been flooding in on us of late about the Universe, humanity can say, as Eddington eloquently says at the end of his remarkable book :—

We have found a strange footprint on the shores of the unknown. We have devised profound theories to account for its origin. At length we have constructed the creature that made the footprint, and lo, the footprint is our own.

Printed and bound by CPI Group (UK) Ltd, Croydon, CR0 4YY

22/10/2024

01777621-0019